THE NEXT
MILLION YEARS

BY

CHARLES GALTON
DARWIN

DOUBLEDAY & COMPANY, INC.

Garden City, New York

1953

Library of Congress Catalog Card Number 52-13372

First published, 1953, in the United States
Copyright, 1952, by Charles Galton Darwin
Printed in the United States

First Edition

PREFACE

WHEN anyone starts to write on a subject, at which he has not hitherto worked professionally, his proper procedure should be to set out on a long course of reading, with careful preparatory annotations of all he has read. Such a course on a tremendous subject like the present one might easily take ten years. At the time when I determined to write this essay I was already over sixty-one, and it is safe to say that it would never have been written, if I had adopted this policy. Since I very much wanted to write it, the only alternative seemed to be to give up the idea of elaborate preparatory reading, and to make use from memory of a very considerable amount of unsystematic reading and thinking on the subject. A book written in this way can of course make no claim to the sort of authority that might be given to one which was based on exhaustive preliminary studies.

I fear that the absence of references will give some inconvenience to my readers. I might be able to quote some of the references, but many of them I could not, and some of these among the most important, so that justice would not be done to the subject by only citing the sources I could recall accurately. In the same way, I have gathered a good many of my ideas from conversations and discussions, in only some of which could I name my informants, so that there again it is juster to name none of them. In view of these doubts about the sources of my knowledge, it would not be proper for me to claim any originality in the views I express; I believe that some of them are original, but even with regard to these I shall not be at all surprised, if it turns out that I have been anticipated.

I have realized to the full the dangers to which I am exposing myself in forgoing the elaborate preliminary studies which the subject demands, but from my experience in other subjects I am encouraged to think that little harm will be done by it. The spirit of criticism is much commoner in the world than the spirit of invention, and progress has often been delayed by authors, who have refused to publish their conclusions until they could feel they had reached a pitch of certainty that was in fact unattainable. Progress in knowledge is more rapidly made by taking the chance of a certain number of errors, since both friends and enemies are only too pleased to exert their critical faculties in pointing out the errors; so they are soon corrected, and little harm is done.

Nevertheless I have taken all possible precautions so as not to make mistakes. I have tried to avoid errors both of principle, and in the examples which I cite, by getting comments from various friends who are well versed in the different branches of the subject, and I have certainly been saved from a good many errors in this way. Lest they should be thought responsible for opinions they may not share, I will preserve their anonymity, but I would take this opportunity to thank them for the great help they have given me.

In composing the essay I have had the difficult task of deciding the degree of knowledge that I might assume in my readers. It has seemed to me wise to err, if anything, on the side of explaining too much rather than too little, and I had therefore better apologize in advance if some readers consider I have wasted their time by explaining in too much detail things with which they were already familiar.

CONTENTS

I

INTRODUCTION

ANYONE who attempts to predict the history of the next ten years is a rash man, and if he attempts to make his forecast for a century he is very properly regarded as so foolhardy as not to be worth listening to at all. Nevertheless, I am proposing here to do what may appear at first sight a very much wilder thing still. I maintain that with our present knowledge of the world and of the things in it, though we cannot at all see the detail, we can foresee the general course its history is almost certain to take over a long period. It is certainly not possible to predict anything like a detailed history of the world, but nevertheless it is now possible to foresee a good deal of what I may call its *average* history. I do not know whether the true historian will admit that this has any claim to the title of history; certainly it could make no claim whatever to be a narrative of events. Its aim is far more modest; it is to describe roughly the kind of things that will be going on most of the time in most places.

In one respect it might appear that the prophet of the next million years had a very easy task. With the great differences of conditions over the different regions of the earth, it is almost inevitable that there must be a great deal of variety in modes of life. In the vast period of a million years, and over the great expanse of the earth, there must be an enormous variety in the happenings, and it may well be so great that, no matter what the prophet foretells, his prophecy will be verified at some place and time. To succeed in prediction in this sort of way would not content me; it would be no better than the prophecies of the fortune-teller, who makes a long list of mutually contradictory statements in the confidence that some at least of them are bound to be verified. My aim on the other hand is to form an estimate of the normal and not the exceptional course of the life of mankind on earth; to describe what will be happening most of the time. It is definitely concerned with the less exciting parts of history, the parts over which the historian's narrative often passes most lightly, because these parts include none of the stirring events which make up the great crises of history.

I need not say that I recognize that this is an exceedingly ambitious programme; probably most readers will start by regarding it as so over-ambitious that it is doomed to failure. My justification for attempting it is that it does appear to me that in the course of the past century, and in particular of quite recent years, there have been such enormous accretions to almost every branch of knowledge, that now as never before an essay like this has become possible. There have been very great historians in past eras, perhaps greater than any we have now, but they simply did not possess the material to accomplish anything of the kind. I have of course no claim whatever to pose as a historian, and it is through my other studies that I have been drawn to this attempt at a synthesis of the various branches of accumulating knowledge in the form of a forecast of future history. I have always had a great interest in history and in the biological sciences, but the final stimulus came from my studies in the physical sciences. It is these that provide the real reason for expecting that something like the present essay may be worth attempting. It may well be that some of my deductions will be corrected by those who have a deeper knowledge of the various subjects than any I can claim. I shall naturally be disappointed if my conclusions have to be corrected or amended in whole or in part, but even if it should be so, I hope I shall have persuaded some readers that the method itself is right; that is to say that it is possible now, with existing knowledge, to make a good forecast of the future fate of the human race.

To justify the principle that we can know what I have called the average history of humanity, I must make a short incursion into physical science. Perhaps I had better begin by reassuring the reader that I shall only need to explain a few generalities, and that no one need fear that there will be much further reference to the subject after the next three or four paragraphs. There are two quite different ways in which inferences, and in particular predictions, are made in physics, and the contrast between them must be made clear.

The older method is the ordinary process of cause and effect. As an example, Newton enunciated the law of universal gravitation, in obedience to which not only the apple falls to earth, but also the moon goes round the earth, and the earth round the sun. This principle enables the astronomer to predict exactly where any planet will appear in the sky at any given future date. Or again take the design of a complicated instrument, such as a television set. The designer arranges his electric circuits and his electronic valves in such a way that, if a specified electro-magnetic signal reaches the antenna, then calculable currents will flow in all the circuits, and these will give rise to calculable streams of electrons in the television tube, which in turn will give a calculable visual image. In this general type of case a definite cause produces a definite effect, and if the effect is not always exactly predictable, that is only because the calculations may be so intricate that the evaluation of the results is not practicable. By those who have not followed recent developments in science, this relation of cause and effect is regarded as the only way by which things can be predicted. For them scientific progress means the discovery of yet more exact effects produced by exact causes, and they conceive that the cause-and-effect relation is the sole idea in what is popularly, if vaguely, called the scientific method.

This would indeed have been broadly true of all the earlier stages of scientific progress and of all the earlier scientific laws that were discovered, but a very different new type of procedure began to emerge some eighty years ago, which has assumed almost dominating importance in recent times. This newer type of reasoning is connected with the principle of probability, and it is unfortunately true that there are a great many people to whom it has not yet become familiar. They find it surprising that the result of a great number of chances may be far more certain than the result of a few. Of course absolute certainty, of the kind given by the cause-and-effect relation, is never attained in this manner, but something effectively just as good does come out of it. As the number of chances becomes larger and larger, the effects of each single event become less and less important, and they tend to cancel out; the probability that they will all go one way becomes quite negligible, so that something approximating with great accuracy towards the average is the final, practically inevitable, consequence.

The classical instance is given by the molecules of a gas. We know a certain amount about the nature of these molecules, and indeed it might be possible at any rate in some of the simpler cases, to work out in detail what happens when two of them collide together, though I cannot call to mind any case where this has actually been done. To do it would be in accordance with the old cause-and-effect physics, but in fact it would not be very useful. This is because the number of molecules is so vast, and their collisions so frequent, that the effect of a single collision is of no interest, but only the average effect of all the collisions. It proves possible to know this average by a very general method, and the average can be found without even invoking many of the properties of the individual molecules, even when these properties are well known. The most general deductions are the *gas laws*, which describe how the pressure of the gas depends on the volume of the containing vessel and the temperature; the most famous is Boyle's Law, which relates pressure to volume. Boyle's Law is verified with the greatest precision and the most absolute regularity whenever it is tested, and yet it is the consequence of the wildly varying and extremely violent collisions between the molecules of the gas. As I have said, in deducing the law it is not even necessary to use all the known properties of the molecules; for example we know with some accuracy the distance between the two atoms of an oxygen molecule, but in fact this distance plays no part whatever in the result. In order to derive Boyle's Law all that is required is the knowledge that the molecules constitute what are technically known as conservative dynamical systems.

For the rest of the gas laws it is true that a little more detail is needed; for example there are differences according to how many atoms there are in each molecule, but a very great deal can be known about the behaviour of the gas with only the sketchiest consideration of the details of the individual molecules. Indeed when a student first encounters the theory leading to the gas laws —it is usually called the theory of statistical mechanics— he is always surprised at the very little foundation that is required in order to establish, quite fully and logically, such an enormous superstructure. Of course it can never be possible to get something out of nothing; there must be a basic principle, and this, as I have said, is the condition that the molecules are conservative dynamical systems. This name is derived from the fact that the total energy of two colliding molecules is conserved, so that it stays constant during the collision, but the term itself has a far deeper significance than this, though it is one which can only be described in technical language. It would involve going rather deeply into the subject to explain it here; and I shall not attempt to do so, since I am only citing it as an analogy, but I must make the point that in this technical language the whole property of conservative dynamical systems can be described in a single sentence. The outcome is that in statistical mechanics, unlike the older cause-and-effect mechanics, the most enormous superstructure can be built, with confidence and certainty, out of a foundation which might appear to be narrow, in the sense that it can be expressed in very few words.

The analogy I have cited of the gas laws is the simplest example that is furnished by statistical mechanics, and it is only fair to mention that, when the subject is pursued further, it does get a good deal more complicated. Thus much greater difficulties arise in considering how the gas can condense into a liquid or solid, but it would not be profitable to follow the analogy into these intricacies. Even in such cases however, though the structure of the molecules must be specified a good deal more fully, the specification still remains fairly simple, and yet it can lead to the most elaborate consequences. The general principle stands, that in statistical theories quite complicated results can be deduced from simple principles.

The internal condition of the gas depends, as I have said, on the molecules being conservative dynamical systems, but there are also external conditions. Boyle's Law relates together the pressure and volume of the gas, so that the measurement of both these quantities must come into the picture somewhere. This they do through

consideration of the containing vessel, for its size determines the volume of the gas, and the pressure is measured
through the force exerted on the vessel's walls. If, then,
predictions are to be made, of the kind yielded by the
methods of statistical mechanics, it is necessary to consider both the internal and the external conditions.

When I claim that we ought to be able to foresee the
general character of the future history of mankind, I am
thinking of this analogy. The operation of the laws of
probability should tend to produce something like certainty. We may, so to speak, reasonably hope to find the
Boyle's Law which controls the behaviour of those very
complicated molecules, the members of the human race,
and from this we should be able to predict something of
man's future. It is not possible to get something out of
nothing here any more than it is in the case of the gas; so
the possibility depends on finding out whether there are
for humanity any similar internal conditions, which
would be analogous to the condition of being a conservative dynamical system, and external conditions
analogous to the containing vessel. If both these demands
can be satisfied, then there is the prospect that a great
deal can be foretold of the future of the human race, and
this without any very close detail in the basic principles
from which it is derived.

Most of the present essay is devoted to discussing the
various principles needed in order to make these predictions, but I will here very briefly anticipate the fuller
discussions that are to follow. In the gas, the external
conditions were given by the containing vessel, and the

analogy here is obviously the earth itself. From geology, we know a great deal about this; for example we know that it has had roughly the same climate for hundreds of millions of years, so that it is nearly certain that the climate will stay the same for one more million years.

The internal principle, which is to be analogous to the property of being conservative dynamical systems, of course lies deeper. It must depend on the laws governing the nature and behaviour of the human molecules. When I compare human beings to molecules, the reader may feel that this is a bad analogy, because unlike a molecule, a man has free will, which makes his actions unpredictable. This is far less important than might appear at first sight, as is witnessed by the very high degree of regularity that is shown by such things as the census returns. When averaged over a whole population these reveal a remarkable degree of regularity in most of the happenings of life; this applies not only to basic things like births and deaths, and uncontrollable things like the incidence of sickness, but also to things in which man does regard himself as an entirely free agent. To take a quite trivial example, there is considerable uniformity in the names given to new-born infants, in the sense that it can be foretold with some accuracy what fraction of them next year will receive each of the more fashionable names. Thus though the individual collisions of the human molecules may be a little less predictable than those of gas molecules—which, as I have said, do not have to be considered in detail either—the census returns show that for a large population the results average out with great accuracy. The internal principle then of the human molecules is human nature itself.

Here once again geology provides help. A study of fossils has shown that it takes roughly a million years of evolution to make a new species of animal; during that time the animal is, it is true, slowly changing, but the cumulative changes are so slow that it is only at the end of that period that the animal can be regarded as sufficiently altered to be dignified by a new name. This principle may be applied to man. For the next million years we shall be concerned with a history governed by the same human nature as we know now, with all its virtues and all its faults. There will, it is true, be small slow changes in human nature as time goes on, but it is only at the end of a million years that it may be expected to have so changed that further prediction about it would become impossible. That is the reason why I have referred to a million years in the title of my essay.

It is worth noting that it is only in very recent times indeed that the present line of argument could possibly have been developed. This is true of every part of the argument. In the first place it is less than a century since anyone realized the compulsive force of the laws of probability, so that before that no one could have conceived that so much could be derived from such a simple foundation. Then again the actual surface of the earth was very incompletely explored until even more recent times, so that something quite unforeseeable might have come out of its unknown regions; the position of the walls of the vessel could not be regarded as fully known. As to the internal condition, human nature, the same is even more true. It is only in the last fifty years that much has been known about the physiology and psychology of the human animal. In particular it is only since then that the principles controlling heredity in animals have begun to be understood, and it is obvious that those principles must govern the characteristics of the human race beyond all else. So until recently there was knowledge neither of the principles, nor of the data, which I am going to attempt to use in the ensuing chapters.

If there were no prospect of deriving anything beyond a sort of flat average of the future of human history, it would be a dull business, but my analogy suggests that it should be possible to carry the matter a good deal further. The physicist can not only give the average state of his gas, but he can also say something about its *fluctuations*, that is to say the manner and the frequency with which it is likely to depart from the average. To explain this, suppose that a sample of the molecules is periodically taken out of one corner of the vessel, say by trapping some of them in a little box. The number of molecules in the box should be in a fixed proportion of the total number in the vessel, and it will never be far from it, but sometimes the sample will have rather too many, and at other times rather too few. Furthermore the sample may vary in other ways; for example it may sometimes be hotter or colder than the gas in the main vessel. These variabilities are what constitute the fluctuations of the gas, and they can actually be observed by suitably designed experiments. They can also be calculated, again without knowing much about the individual molecules, so that it is possible to say what variations will occur, and how often they are likely to recur, though it is not at all possible to predict when they will occur. In the same way it should be possible to get some idea of the fluctuations of human history, which will lend a variety to the dead level of the average. Indeed these fluctuations will be far more prominent in human history than are those which are observed in a gas, for the reason that, proportionately speaking, fluctuations decrease as the total number of molecules increases. Now in the course of a million years there will have been a very great number of human beings, but even so it is incomparably fewer than the number of molecules in even quite a small vessel. So the fluctuations in human history will be far more prominent than the fluctuations in a gas.

Much of what will happen in the future can be read from consideration of the past, but there has on the whole been a tendency for historians, apart from their primary function of recording the past, to be interested not so much in resemblances as in differences. They have tended to emphasize the difference of structure of the civilizations of Egypt and Babylon, rather than to point out how much more alike they were, than was either to the contemporary barbarism of Europe. A distinguished exception to this has been the great work of Arnold Toynbee, who has studied what might be called the Natural History of Civilizations. This is an incomparably finer project than any I am competent to attempt, but it does adopt a similar line of thought, that of determining whether laws can be laid down from which the march of humanity can be foreseen.

The plan of the present work follows in a general way the course suggested by the analogy I have been drawing from physics, though there will be no further reference to this source. In the next chapter the subject is population. This is the most fundamental question of all, for if there were no human beings there could be no human history. Then there are two chapters dealing with what I have called the external conditions. One of these briefly reviews past history; its aim is to mention only the fundamentally important things, those that one might imagine would be noticed by an historian of a million years hence. The other deals with the physical conditions that may be expected to prevail in the world. Then there are three chapters dealing with what in my analogy I have called the internal conditions. These contain a discussion of what appear to be the chief qualities of the human animal, in so far as those qualities are likely to help his survival in the struggle for life. In the first of them he is regarded as a wild animal just like any other animal, and I consider the qualities which will help him from that point of view. The next chapter is concerned with the influence of his social qualities, in which he is so different from other animals. The third is devoted to settling the balance between the influences discussed in the previous two chapters, and in particular to the consideration of whether man is a wild or a tame animal. After this there are two chapters dealing with matters of a slightly less fundamental character, though they are much too important to be omitted. The first examines the effect of the limitation of populations, such as that which is being experienced at the present time in many of the countries of western Europe. The second considers the interesting, if less important, question whether man, in his perpetual striving after happiness, is at all likely to achieve it. In the final chapter I attempt a synthesis of all these things in the form of a forecast of the history of the future. It is divided into several sub-sections, each of which deals with one of the main aspects of human life.

I have attempted to arrange the whole essay on the plan that all the earlier chapters should lay down principles, and that the final chapter should contain their applications to the actual history. In laying the principles down however, it has been necessary to present a good deal of detail in order to explain and to illustrate them, so that in fact a good many of the applications will be found in the earlier chapters. I fear that this may be inconvenient in some cases, because such illustrations often have implications over a wider field than the principle immediately concerned, and thus the reader may be left with the feeling that the discussion is sometimes incomplete. In the final chapter therefore I have attempted to gather these threads together, as well as to fulfil my main aim of making the forecast of the general history of the future.

Before developing my arguments, it may be well to warn the reader that the consequences I am forced to deduce will be found exceedingly depressing by all the political and social standards that are now current. It will not perhaps be quite true that history will be "nothing but a record of the crimes and follies of humanity", but it will be much more like that than like the utopias—for the most part, be it confessed, the rather depressing utopias—which have been expounded by all the idealists. If the world is inevitably to be so much more dreadful a place than current thought expects, would it not perhaps be better to forget the fact and simply go on hoping? I do not think so; if we are living in a fool's paradise, it is surely better to know the fact. But the matter goes further than this; for we certainly can do something to control the world around us, and if we can appreciate the limits of what is possible, we may have some hope of achieving our aims, whereas if our aims are outside possibility, then we are doomed to failure. Therefore it is a practically important thing to see clearly any laws which must set absolute limits to what it is possible to do.

A parallel will make this clearer. In the eighteenth century the state of knowledge of mechanics was very imperfect and many inventors devoted their time to trying to invent ways of making engines, so as to provide power for machinery. There was no known limit to what could be attained. Even though it had already been adumbrated that the *perpetuum mobile* was not possible, the principle was not understood at all exactly, and an inventor might not unreasonably at that time have felt that there was no limit at all to what he might hope to achieve in the way of inventing an engine. Almost exactly a hundred years ago the situation changed completely when the laws of thermodynamics were formulated; these laws set very precise and absolute limits on what is attainable in an engine. At that stage the optimistic inventor, untrained in scientific principles, may have felt that the world had collapsed; what was the use of his going on in a world lost to all ideals, a world which insisted on a dreary uniformity with no hope that mankind would rise superior to the deadening requirements of the laws of mechanics? For him it must have been a bad world. Not so for the true engineer. He now could know just what was physically possible, and could set himself a target that was actually attainable with the sure knowledge that he might achieve something towards it. Through the recognition of these absolute limitations on what he can do the engineer now has a better, not a worse, prospect of doing good.

Can we not draw from this parallel the conclusion that if we know the limit of what is possible for humanity, through determining some kind of laws of human thermodynamics, we shall be more successful in doing good in the world, than if we recognize no limitations, and so are perpetually struggling to achieve what is in fact quite impossible? I am going to try to see what some of these laws of human thermodynamics are; of course they cannot be expected to have the hard outline of the laws of physical science, but still I think some of them can be given a fairly definite form. It will be for others more skilled in biology than I can claim to be, to perfect, or perhaps to correct, these attempts that I shall be making.

II

POPULATION

THE central guiding theme that must run through all considerations of history is the question of population, and there is a difference here between future history and past history. In past history the people were there and the historian watches what they did; in future history he has to start with the more basic question of what people will be there. The question as to which peoples survive in the world during the march of the ages is fundamental, and must override all questions as to whether future man will be better or worse than present man, or whether he will rise to heights we cannot conceive or sink to levels we should despise. The fundamental question is survival, and this must never be forgotten, but by itself it is entirely unsatisfying, because we do want to make judgments of quality about our descendants. In this chapter then I shall first of all deal with general questions about population, but towards its end I shall turn to some of these interesting questions of quality. It must be insisted, however, that though these have much more appeal than has the fact of survival, they are secondary to it, because it is only the races that survive that make the history.

It is always necessary—and it is indeed quite surprisingly difficult—to keep in mind that the fundamental quality pertaining to man is not that he should be good or bad, wise or stupid, but merely that he should be alive and not dead. Therefore the first thing that must be asked about future man is whether he will be alive, and will know how to keep alive, and not whether it is a good thing that he should be alive. Whether we like this fact or not, it has the advantage of providing an *objective* basis for the study of future history, which simplifies things a great deal because it eliminates some of the danger of our forming prejudiced judgments. There may be endless arguments as to which of the individuals A or B is the more estimable member of a community; tastes differ and agreement may never be reached. But there is a far better chance of reaching agreement on the brute question of which of the two is likely to survive, whether it is in person, or through his offspring, or by creating a successful polity for his community. These are objective judgments, and so they are likely to be free—or anyhow much more nearly free —from the prejudices, which none of us can help having when it is a matter of making subjective judgments about human values.

The primary question then arises: what are the conditions which determine whether a man will survive or not? In the case of animals, there is a great variety in the threats that determine their survival. Some are attacked by beasts of prey, some by parasites, some by pestilences, while some whole races of animals have been destroyed by catastrophes, such as the submergence of land under the sea; but all without exception are subject to one overriding condition, the danger that they may not get enough to eat. This gives rise to yet another threat to the survival of the individual animal, the competition between the different members of the same species for limited supplies of food. Man can rise superior to most of the threats that affect the rest of the animal kingdom; he can dominate the largest and fiercest beasts of prey, he has already learnt how to control most of his parasites, and through medical science he is even learning how to control the most deadly of all, the bacteria of disease; while if his land is drowned under the sea he can take ship and sail away. There remains for him one condition that he still shares with the rest of the animal kingdom as a perpetual menace to his life, and this is the need of food. Here again there arises the competition for limited supplies, whether it is a competition between man and man, or between nation and nation, and for humanity this competition assumes far greater importance than it does for any of the animals, just because, unlike them, he has succeeded in overcoming so many of his other enemies. It is this competition that will determine the detail of history, in the sense that it will determine which men and which races will survive; but deeper than this there lies the question of how the survivors are to keep alive, and the final controlling condition for this is their supply of food. It is food that in the end determines the population of the world.

During the past century many writers have discussed the question of population, and they have naturally been chiefly concerned with the conditions of the present, of the recent past and of the immediate future. Here however, I am not concerned with a century or two, but with a million years, and for this it will suffice to go back to the founder of the study of population. A hundred and fifty years ago Malthus wrote his *Essay on Population* in which he drew attention to the conflict between the law of biological increase of the human species which is a geometrical progression, and the law of increase in the area under agriculture which can only, roughly speaking, be an arithmetical progression. Man must always be outrunning his food supplies. Malthus himself, and others after him, tried to devise ways of escape from this threat, but it has never been really disposed of, and it has only escaped the predominating attention it deserved through the accident of the history of the nineteenth century. It was verified that the increase in population tended during that period to be in geometrical progression, but the development of the New World, and the establishment of railways and steamships to carry its products to the Old World had the unforeseen consequence that for the best part of a century the cultivated areas could increase at a rate greater than the population. Malthus's first principle was shown to be correct, but his second was vitiated by the quite exceptional conditions of the nineteenth century. This era is probably now nearing its end, and the difficulties he expounded must be faced.

There are no doubt many who are not familiar with the argument of Malthus, and so it may be well to describe how it works out numerically. His first hypothesis is that there is a natural rate of increase for any species of animal, and if one thinks for example of a cow producing one calf every year for say five years, it seems a very reasonable hypothesis. The rate certainly varies very much from one species to another, and the increase is restrained in nature by all sorts of checks, the chief of which is shortage of food, but Malthus assumes that for any species there is a natural rate of increase which would operate in the absence of these checks. We can estimate fairly well what this rate is for humanity by the experience of the last century in these islands, for during that time the main checks on natural increase were removed through the importation of food and improved sanitation, and the population was quadrupled in the century. There have been corresponding increases in many other countries of Europe, Asia and America, but not all on quite the same scale, so I will take a cautious estimate and assume that the natural rate of increase for man is that he should double his numbers in a century. It may be mentioned that these estimates are all well below some of the values that Malthus himself quotes.

Now look at the other side of the account. The present food production of the world is roughly about enough to feed the population of the world; this is almost a truism, for if there were not enough, the excess population would have to die, and, if there were too much, the excess food would simply be wasted. If the natural increase of population is to be met, more food will have to be produced, and this can be done to some extent by improved farming and by bringing more land into cultivation. There is nothing unreasonable in saying

C

that the food production of the world could be doubled or trebled; but it is rather hard to see how it could be raised more than ten times on the present methods of agriculture. But it is very possible that these methods could be improved out of all knowledge, and for the present argument I am quite ready to grant that the food production of the world could be increased a thousand times above its present level, even though I do not believe it possible. But there is no need to stop even there, for there is the vast area of the ocean, which we hardly exploit at all at present. Here again I do not believe that any enormous increase could come out of the intensive cultivation of the sea, but for the sake of the argument I am ready to grant it, and, to take a figure beyond what anyone is likely to think possible, I will assume that the total food production of our planet might be a million times what it is now.

Now bring these two sets of figures together. If the population is doubled in a century, it is only three and a half centuries before it will be ten times its present number, and this would exhaust what I have estimated as the possibilities of the existing systems of agriculture. Keeping on at the same natural rate of increase, the population will have increased a thousandfold* in ten centuries, and even if new agricultural methods should permit the production of a thousand times as much food as at present, there would by then still only be just enough food to support the population. And a thousand years is

* The arithmetic of these calculations is simplified by noting that doubling a number ten times multiplies it by 1024, which is roughly a thousand.

a short period even in the span of known past history, and quite insignificant when counted on the scale of a million years. Again, a population, that has a natural rate of increase of a thousandfold in a thousand years, will increase a millionfold in two thousand years, and so at the end of that two thousand years there would be need of the enormous quantity of food of a million times the present amount. It is evident that no increase of food production, however fantastically imagined, could cope with the natural increase of mankind for more than a very small fraction of a million years.

The whole argument is hardly affected even if the natural rate of increase has been much over-estimated. Though experience is all against it, suppose that the natural increase of mankind would double the population in a thousand years instead of a hundred. The only effect would be that it would take twenty thousand years instead of two thousand for the population to multiply itself by a million, and twenty thousand years is still a very short span compared to a million years. All these figures illustrate the general principle, familiar to the mathematician, which may be expressed colloquially by saying that it is quite impossible for any arithmetical progression to fight against a geometrical progression.

To summarize the Malthusian doctrine, there can never be more people than there is food for. There will not be less, because man, like every other animal, tends to increase in numbers. There have been a few exceptions to the rule of man's natural increase, and a most important one, which will be examined later, is present with us now, but to a quite preponderating extent the rule has held in the past, and there is every likelihood that it will continue to hold. The straightforward way of striking the balance is nature's method of creating an excess and then killing it off by plague or starvation. Malthus himself, and other more recent writers also, have attempted to propose solutions which should allow us to escape from this threat, but nobody has found one which is at all convincing. It follows that in the very long run of a million years the general course of future history is most of the time likely to be what it has been for most of past time, a continual pressure of population on its means of subsistence, with a margin of the population unable to survive.

There is no escape from the fact of the finiteness of the amount of food that the earth could produce; but Malthus's first hypothesis, that there is a natural rate of increase of man, is much more likely to be questioned. It may be an over-simplification to take the rate as fixed, but it is indisputable that animals, given favourable conditions, do rapidly multiply to fill the vacant spaces of the world. The same has been true of the rather slow-breeding animal man, and it has been confirmed by the last century, when for a time the threat of food shortage was removed in some countries. Indeed this period has had the curious consequence of allowing people to forget Malthus altogether, since the increase in agriculture outstripped the human rate of increase for a time, and so drew attention away from the problem of population.

Our insight into the matter has further been confused by the fact that at the present time we are threatened with a decreasing population in England, and indeed in many of the countries inhabited by the white races; this is a very important phenomenon indeed, and it seems to contradict Malthus's principles. It will be the subject of a later chapter, but the matter must be regarded on a world-wide basis, and not just as one of Western Europe or North America, and anyone who has, for example, visited India will get a very different impression and one which is juster. Thus not long ago the province of Sind was mainly desert; the ground was quite fertile but there was no rainfall. A great engineering undertaking, the Sukkur barrage, has spread the waters of the Indus over a very wide area, and turned much of the desert into a garden. According to the universally accepted standards this was a great benefit to the world, for it made possible the adequate feeding of a people previously on the verge of starvation. But things did not work out like that, for after a few years the effect was only to have a large number of people on the verge of starvation instead of a small number. This is not the place to raise the moral issue of whether the world is the better for having the Sukkur barrage or not; from the point of view of population it has had the effect of increasing somewhat the already great importance of the contribution of India to the population of the world.

It must be accepted that the objective fact of survival is more fundamental than any question of the quality of the surviving life, good or bad, and this consideration gives a colour to some of the happenings of past history which is rather different from the colour in which they have often been presented. To illustrate the point I will take an example. We are all shocked when we read accounts of child labour in the factories of the early nineteenth century, and we can all agree that the conditions in many of the factories were terrible. But how did it come about that, as soon as there were factories needing labour, the children were there to undertake it? The most reasonable explanation is that in the previous generations most of the children simply had to die in infancy, and that it was the factories that saved the lives of the new generations. For long ages the world had got used to a very high death rate of infants, and took it for granted that this was an inevitable law, and now suddenly it was found that the law was not inevitable, and that the infants did not have to die. It was the factories that saved all these lives, all too many it is true for only a few years; but still many did grow up, and since it is life and not no-life that counts, the factories might claim to be benefiting the world. In saying this I do not of course in the least want to condone the system, which sometimes exhibited a monstrous cruelty on the part of selfish employers, who were enriching themselves at the expense of the unnecessary sufferings of their fellow creatures. Still, in weighing the question up, there should be counted on the positive side the fact that quite a large fraction of our present population would simply not be in existence at all now, if there had been no factories a hundred years ago.

I have already pointed out that though the availability

of food is the fundamental question for mankind, there is also the important question of the competition for that food between men and between nations. It will be those who are successful in the competition who will make up the population of the future, and so it is the qualities that lead to this success which will determine the course of future history. The consideration of these qualities is therefore naturally the main theme of the present work, and they will be studied in the ensuing chapters. In the remainder of this chapter I shall try to deal with certain arguments about population, which I suspect may be present in the minds of some of my readers. In looking at past history they may have been accustomed to consider that one of the important things to do—as it is certainly the interesting thing to do—is to assess merits in the personalities of their histories, and they will not be content to believe that a cold counting of heads is really more important. I do not in the least want to oppose the making of judgments of this kind, and indeed I shall be making many myself, but here I want to establish the point that, just because they are judgments of the past and not the future, many of them are irrelevant to the subject.

Most people are much more interested in quality than in quantity, and they may argue that there have been many cases where quality has proved itself more important than quantity. They may say that in the course of past history a numerically small race of high quality has often been far more important than a large race of low quality. This has been true in the past, and no doubt it will often be true again, but taken by itself the judgment accomplishes nothing. When it is said that a small race was often more important than a large one, it sometimes means that this race, in consequence of its high quality, achieved success in life in such a way as to become fruitful and multiply, so that in fact it later became a large race. Rome in its early days is a typical example; it is not important because of its smallness—for that would imply that the rival city of Veii was equally important—but because it ultimately became large. We do not especially admire each of the villages of Latium just because one of them grew into an empire, and we do in fact value the little Rome only because it became the great Rome. In the march of history every institution has a small beginning, but it is the whole of its history and not the beginning that must count in the assessment of its value, so that in such examples as this it is irrelevant to emphasize the smallness.

There is a second very different sense in which it may be said that there are numerically small races which are more important than large ones. For example there is the group of a few among the Athenians of the classical period who made important contributions to knowledge and art. The number concerned was extraordinarily small, and their era was very ephemeral, but their enormous contribution to the richness of the world is indisputable. Such contributions are undoubtedly among the most important things in the world, but they are nearly irrelevant in the present context. There have been innumerable small city states, whose earlier histories were indistinguishable from that of Athens, and it is only after the event that we can discriminate Athens from them. For future history our enthusiasm cannot be expended on all these innumerable little states; we should only be justified in doing so, if we could hope thereby consciously to create something like a new Athens.

A study of past history does not encourage this hope. Most of these flowerings have occurred in association with the rather sudden acquisition of wealth by a race, wealth often won by the conquest or the commercial exploitation of neighbours. But the converse has not been true, since such wealth has frequently been gained without any efflorescence of the arts or sciences. In fact these efflorescences are what by my analogy from physics I have called fluctuations, representing occasional extreme departures from the average. If it is going to be hard to do anything in the way of controlling the average history of humanity, it is going *a fortiori* to be very much harder to control its fluctuations. To indulge in a flight of fancy, imagine that a world dictator considered that the only really important thing was to have a new school of painting as great as the Italian or Dutch schools. How should he go about creating it? To judge by past history he would not succeed by founding learned colleges of art with elaborate provisions for competitive scholarships, but rather by creating a thoroughly turbulent world, full of struggle, warfare and injustice. In this world here and there cities or countries would arise which through the ability of a few of their citizens, their Medicis or their Amsterdam merchants, attained a very unequal share of the world's wealth. By the time two or three dozen

41

states of this kind had come into existence there might be a faint hope that in one or two of them there really would have arisen simultaneously patrons with the taste of a Lorenzo and painters with the genius of a Titian or a Rembrandt. Altogether it does not seem likely that the world dictator would be very successful. If this is really the most important thing for the world, it does not seem likely that we can do much to bring it about.

These examples from the past, where we can be wise after the event, give little help in suggesting how the qualities of the population of the future are to be judged. Survival is the essential factor in the making of history, and it must certainly have first place, but most of us want to know much more than this about the qualities of the survivors. What line should the historian take in making judgments about these qualities? It would be a tenable view that his duty is coldly and objectively to observe what happens, noting merely that such and such a population flourished at such and such an epoch, and holding that it was not for him to comment, either favourably or unfavourably, on the qualities of the population. If the future is regarded as a quite uncontrollable unfolding of events, then a cold account of it, free from all moral judgments would be an admissible policy. The only important thing in such a view would be the purely objective question what at some epoch the surviving population was and what was the character of its life, even if it had degenerated to something very much lower in the animal scale than anything we have at present. But even for the immutable past most historians do not follow this method; they do pronounce judgments, though nothing they say can alter what actually happened, and most of us accept this very definitely as the best way to write history. If that is so for the past, how much more is it so for the future, for though our control of the future is certainly very much smaller than is claimed by the optimists, still some control does exist; to some extent we may aspire to give a direction to the development of the world.

Historians of the past have usually taken some broad idea as a guiding principle in their account of past events, and there has been a good deal of variety in these ideas. To one it will have been the material conditions of a people that is of chief importance, to another their political institutions, a third will be inspired by their philosophic or religious thought, and a fourth by their military exploits. Another will trace the history of a broad general idea like the development of personal liberty, while others have, perhaps unconsciously, imbibed the tenets of some long-dead narrow political party, and have judged the events and the personalities of their historical period by that standard. Every historian must be allowed to have some guiding principle of this type as a background for his history, and I am entitled to claim this right for myself. Since I have been emphasizing the fundamental position of the question of survival, it might be considered that I ought to refrain from all intellectual and moral judgments about the future members of the human race. It is simply impossible for any human being to pursue such a course, because his whole life has been coloured and conditioned by the habit of forming judgments of this kind. Even if I tried to do it, it would not be possible for that reason; but, guided by the example of the historians of the past, I would not wish to do it.

Though the matter of survival is fundamental, still it is permissible to show preferences between different ways of surviving. For example some highly successful modes of life—such as that of the parasite—would not be regarded as admirable, no matter what human standard they are judged by. Now the chief natural qualities of man, which distinguish him from other animals, are that he is simultaneously an intelligent and a social animal, and both these qualities tend towards success in survival, the one for the individual, the other for his tribe. Both are qualities which are admired at any rate by the majority of us, some putting intelligence first, others the sense of social duty. Therefore in so far as it is possible to look beyond the brute question of survival and to make subjective estimates of value about the future human race, I shall rate as admirable any improvement that in the course of the ages should develop in the intellect of mankind, and any improvement in his sense of devotion to his fellow man. A combination of the two qualities is best of all, but if it is necessary to select between them, I should assign first place to intelligence, if only because it is a more distinctive characteristic of the human race than the social sense, which after all man shares with many other animals. In studying past history it is only possible for the historian to take what did happen and either approve or disapprove. In future history the historian is not so limited; he may not only approve or disapprove, but he may also hope. I shall hope most of all that the surviving races of man in the long ages to come will increase still further in intellectual stature.

III

THE FOUR REVOLUTIONS

TO set the stage for the history of the future, it is natural to start by reviewing the history of the past, having regard only to the principal facts. I am trying to imagine what an historian of a million years hence, engaged in preparing a universal history of the human race, would select from our own past history as worthy of notice. I think he would select only occurrences where mankind made a step forward which was never lost again; they might be called the *irreversible* stages. There would seem to have been four such stages in the development of humanity, since the time when *homo sapiens* came into existence. With the first three everyone is familiar, so that they need only be touched on, but the fourth, which is quite as important, is so recent that it has almost escaped conscious notice. I shall call each of them a revolution, though the word is not meant to imply any extreme suddenness. In each, the germs may be detected long before, and it may have been a long time before they spread over the world; in some cases the revolution has been made independently in different regions. The central feature of each revolution has been to make it possible for mankind largely to multiply in numbers.

The first revolution occurred long before the dawn

46

of history, and we can only conjecture its effects, though we can do this with confidence. It is the discovery of fire. By means of fire cooking becomes possible, and so the difficulties of man through his extremely poor equipment of teeth can be overcome. The possibilities of diet are multiplied very many times, both because meat can be eaten that is not completely putrified, and because many herbs thereby become digestible and nourishing. It can confidently be said that as soon as fire came into use, the earth could support a much increased population, because so many more varieties of food became available. There was of course also a second use for fire in the heating of shelters, which was important though by no means so important, since man could thereby live in the temperate and sub-arctic regions, in a way that would not have been otherwise possible.

The second revolution is the invention of agriculture. This dates from the neolithic period, perhaps ten or fifteen thousand years ago, so that a good deal is known about it. The tribes that had agriculture could provide themselves with food, both animal and vegetable, far more regularly than was ever possible for the hunters or food-seekers. They would become much more free than the hunters from the difficulties of the seasonal cycle, and could settle permanently in one place in much larger communities. Once again, with the invention of agriculture, there must have been a great expansion of populations.

The third revolution is the urban revolution, the invention of living in cities. This revolution arose in several different places, at different times and apparently independently; the chief places would perhaps be Egypt, Iraq, China, Mexico, and the earliest time was about six thousand years ago. By the close association in cities, bringing with it the division of labour, the establishment of food stores and the possibility of relieving local shortages through the regular operation of trade, it once again became possible greatly to increase the population. All this is of course in the historical period, and a great deal is known about it, so much indeed that there has been a tendency to study and to emphasize the differences between the various civilizations, rather than their resemblances. In getting a true perspective of the world, it is more important to remember that life in Egypt and life in China were far more alike than either was to life at the same period in Europe. On the analogy I made between human history and the molecules of a gas, the different civilizations are to be ranked as fluctuations from the average; they have gone in rather varied directions with most interesting differences, but it is far more fundamentally important to notice not these differences but the resemblances.

The fourth revolution in human history is so recent that it has hardly been recognized, because we are still in the middle of it, so that we lack the perspective to compare it with the others. It may be called the Scientific Revolution, for it is based on the discovery that it is possible consciously to make discoveries about the fundamental nature of the world, so that by their means man can intentionally and deliberately alter his way of life. Our histories are so detailed, and run so uniformly through this period, that it has hardly been noticed as constituting a revolution. That it is so may be perceived by observing that the population of Britain has increased more than four-fold since 1800, and much the same is true of many other parts of the world, by no means exclusively among the white races. Moreover during the last one hundred and fifty years the whole manner of living has been more changed than in the previous fifteen hundred years. It is true that life in western Europe in 1750 was very materially different from life in A.D. 100 in Italy; Gibbon notes, almost with surprise, that at its zenith the population of Rome was considerably smaller than that of London in his own day. Since London, unlike Rome, was by no means unique among cities, this shows a considerable advance in the art of living close together on the ground, but it is likely that it was due to steady, though not revolutionary, improvements in transportation, in particular water transport, since this would make very much easier the transport of food into concentrated areas. In other matters, too, of course, there were important changes, such as printing and the military arts consequent on the use of gunpowder, but without belittling these changes, they were on an incomparably smaller scale than those witnessed between 1750 and 1950, in nearly all parts of the world. It would surely be just to say that London in 1750 was far more like Rome in A.D. 100, than like either London or Rome in 1950.

Germs of the scientific revolution can of course be

seen long before its actual birth, just as no doubt there was sporadic agriculture before the neolithic revolution. There were discoveries, and very useful discoveries— then as now unfortunately it was military science that seems to have progressed most—but there appears to have been little idea that discoveries or inventions could be deliberately made of such a character as really to alter the world. The germinating idea is to be found in the experiments of Galileo and in the writings of Bacon, but the revolution may be said to have been born at the time of the English Industrial Revolution, and in particular through the invention of railways. In this revolution, unlike the previous ones, we can have an exact knowledge of the effect on population. In a century the population of England, in spite of much emigration, was multiplied by four, and this alone shows what an exceptional period it has been, for, if the same factor of multiplication were to continue the result would give a quite fantastically impossible increment in even a thousand years. The principal contributions to this revolution have come from the Atlantic sea-board, and the greatest increments have therefore been among the white races, but the benefits have been shared by most other parts of the world. For example the population of India used to be held in check by periodic famines and pestilences, but the introduction of modern hygiene, and the administration of the famine code, made possible by railway communications, have had the effect of increasing the population of India, at a guess by a factor of more than two in a century.

The central fact of this revolution has been the discovery that nature can be controlled and conditions modified intentionally, but just as the city may be regarded as the symbol of the urban revolution, so there is a symbol for the new one. It is the fact that the earth has become finite. There are no longer any frontiers containing the unknown, and nothing can happen in any part of the world which may not have important effects anywhere else. In the long past there was always the danger of invasion from beyond the frontier by the army of some unknown and perhaps superior civilization, or conversely there was the possibility of the colonization of a vast fertile unoccupied country. There is still a very great uncertainty about what incursions there may be from other parts of the world, but the uncertainty is now one about human nature, no longer about geography. The finiteness of the world is one of the chief things that make it possible to foresee its future with a degree of confidence now that would have been impossible little more than a century ago.

The word *civilization* signifies in its origin the mode of life connected with living in a city, and, since there has been a great variety in the modes of life practised in the different cities, it is reasonable to speak of many different civilizations. There are still very different modes of life in different parts of the world, but they are united throughout the whole world by the new knowledge and the new mode introduced by the scientific revolution. There is really need for another word to replace the word civilization, a word which would connote the universality of the new culture, but no such word has come into use, and I shall not attempt to invent one. If there were such a word, it would be accurate, and not cheaply cynical, to say that the fourth revolution has destroyed civilization, for it has replaced it by the new and superior mode of life.

It is a natural question to ask whether there may not be other revolutions in store for humanity. The answer is that one future revolution is nearly a certainty, while there may well be others. The fifth revolution will come when we have spent the stores of coal and oil that have been accumulating in the earth during hundreds of millions of years. This will probably be well within a thousand years, a very much shorter period than the periods between the other revolutions. It is to be hoped that before then other sources of energy will have been developed; the subject will be discussed in detail in the next chapter, but without considering the detail it is obvious that there will be a very great difference in ways of life. After all a man has to alter his way of life considerably when, after living for years on his capital, he suddenly finds he has to earn any money he wants to spend. Whether a convenient substitute for the present fuels is found or not, there can be no doubt that there will have to be a great change in ways of life. This change may justly be called a revolution, but it differs from all the preceding ones in that there is no likelihood of its leading to increases of population, but even perhaps to the reverse.

What other revolutions in the remoter future may there be in store? This is perhaps not a very profitable speculation, since such revolutions cannot be foreseen;

if any one of them could, we should by the very fact be on the highway towards it. Nevertheless, with this caution in mind, it is interesting to make conjectures about the subject, bringing to bear on it the very considerable knowledge we now possess of the nature of the world round us, and allowing free rein to the imagination. There will no doubt be periods again, when the world flourishes exceedingly, but these do not necessarily count as revolutions. The essential feature of a revolution is that there should be an *irreversible* change in the ways of life. Thus suppose that after the practice of agriculture had been well established, there had arisen a reversion from it over a great part of the earth; in consequence of this so many people would have had to die, that the remainder would certainly have seen the error of their ways and returned to the practice. The four past revolutions have all had this quality of irreversibility, and so has the fifth which I have adumbrated. All these revolutions have been concerned with man's control over his external surroundings, and my first speculations will be as to possible extensions of this kind.

It must first be recognized that the impulse of the fourth revolution is by no means exhausted yet. Even without any new discoveries at all—and new discoveries are being made every year—there would be very great changes still to come in the near future. The population of the earth could increase very greatly without any new discoveries at all; it might reach a level of density over many parts of the earth as great as it is now in the more populous regions. Even if this expansion were to take several centuries to come about, it should be counted as belonging to the fourth revolution. Future scientific discoveries may lead to other advances, but these could only be counted as belonging to a new and separate revolution, if the present series of advances came to an end and was followed by a period of comparative stagnation for a few thousand years.

The most likely cause of another revolution would be the discovery of some new large source of human food. It might be found possible to synthesize food from its chemical elements, or it might, for example, be found possible to turn grass or wood into a satisfactory human diet. This would constitute a new revolution, the second scientific revolution; for the new food supplies would induce an enormous increase in population, and once the practice had become widespread, there could be no drawing back. I shall not consider the matter further here, since the question of food supplies is reviewed in a later chapter.

Could there be any discovery in the arts, as opposed to the sciences, that might lead to a revolution? There will undoubtedly be many new and exciting discoveries in the arts; there will be new schools of painting, music and literature, and these will contribute a great deal to the happiness of the world, or at any rate of a great many people in it. But they hardly seem to fall into the class of what I have called revolutions, for it does not seem that through them any radical change could arise, which would irreversibly alter the ways of life of hundreds of millions of the human race. It seems precisely in this condition of irreversibility that the arts fail, for in them, much more than in other branches of knowledge, there is a frequent tendency to revert to earlier models; in this sense reversibility is an important characteristic of the arts.

There is another imaginable revolution which would occur if, by any means whatever, it were found possible to foresee the future with substantially greater accuracy than we now can, so that it might become possible to know with a good deal of confidence the most probable consequences of any proposed plan of action. It is what we all try to do even now as far as we can—it is indeed what I am trying to do in the present essay on the long-term scale—but I am imagining that some new discovery should make the process far more precise for short-term planning. This might come about, for example, through the use of the new high-speed calculating machines, which in a short space of time might explore the consequences of alternative policies with a completeness that is far beyond anything that the human mind can aspire to achieve directly. If this were the way the revolution was made, it would have to count as yet another scientific revolution. But I do not want to exclude the possibility that it might all come about by some other non-scientific means, though it seems a good deal less probable that this should be so. I do not believe that the Delphic oracle will be revived, and if future famines are to be foreseen and avoided, it is far more likely that it will be done by scientific weather forecasting than through Joseph's interpretation of Pharaoh's dream. However this may be, if the future could be more confidently predicted, it would evidently have an immense effect on world history. For example, no country would embark on a plan of rapid world conquest, if it could foresee that the war would almost certainly end in crushing defeat after six years. The possibility of making such predictions would have the real character of an irreversible revolution, in the sense that no nation which had grown used to consulting the new—and reliable—augurs, would ever revert to the haphazard methods, which are all that we possess at present.

These possible revolutions share with the past revolutions the quality that they would increase man's control over external nature; the fifth revolution, the shortage of fuel which I have adumbrated, will in fact decrease it, but it will have the same character of being an external revolution. But there is also the possibility of an internal revolution. This would come about if means were discovered of deliberately altering human nature itself. I shall discuss this in later chapters after a closer review of the innate qualities of mankind; here it must suffice to say that the prospects do not seem at all good. There is first the extreme difficulty of making such changes, and the probability that most of them would be for the worse, and secondly, if by chance a revolutionary improvement should arise, it seems all too likely that the rest of mankind would not tolerate the supermen and would destroy them before ever they had the time to multiply. It was mainly the belief that there will be no revolutionary change in human nature, that emboldened me to write this essay.

As I have said, these speculations about future revolutions are only the wildest conjectures. Leaving aside the unknown date of the fire revolution, we know that within less than the past twenty thousand years there have already been three revolutions. If this can be regarded as a precedent it suggests that there should be a revolution at least every ten thousand years; that is to say more than a hundred of them in the span of a million years. I confess to very considerable doubt as to the likelihood that there are so many revolutions in store for our descendants, but at any rate it is hardly profitable to speculate further on the subject.

IV

MATERIAL CONDITIONS

THE future of the human race of course must depend on the nature of the inorganic world in which it lives, so that it is well to begin by reviewing this. In the first place all astronomical and geological evidence indicates that the climate of the earth has been roughly constant for more than a thousand million years, and there is every reason to think it will continue so for many million years to come. There is always of course the chance that there may be a dark star moving through space towards the solar system so as to collide with it. The collision need not be very severe to end the history of the human race, for a perturbation of the earth's orbit, which might from the astronomical point of view be counted as quite small, would be sufficient to change the climate enough to destroy all life.

We obviously cannot know whether there is a dark star approaching us, because it would be invisible until it was quite near, but we can say that it is extremely improbable. First, if there were many such stars, one of them would probably already have hit the solar system during the era of two thousand million years for which the earth has existed. Secondly, in the intensive study of the heavens by astronomers, collisions would have been observed between other stars, and though new stars, *novae*, are found rather frequently, their character does not suggest that they were caused by collisions of this kind. There is also another class of new stars, the *super-novae*, only rather recently recognized; the last one that occurred in the galaxy was Tycho's star, which happened in 1572—for some time it was so bright as to be visible in daylight. It is still very doubtful what makes a supernova; it might be a stage in the life of every star, but their rarity makes this unlikely, and as the sun is by all standards a very normal, astrophysically uninteresting star, we can be fairly sure that it will not blow up in this way. The general conclusion of the astronomical evidence is that it is very unlikely indeed that there should be a catastrophic end to the earth in a million years, or any substantial change in its condition.

Though the earth's climate has been roughly constant for so long, there have been minor fluctuations in it. Thus in England we are only now emerging from an ice age. This is the last of four recent periods of glaciation in the northern hemisphere, and there were three intermissions between these periods when the climate was even warmer than it now is for quite a long time. We cannot therefore be sure that there are no further ice ages coming to us. All that can be said is that though there have undoubtedly been other ice ages in the more distant past, they are geologically speaking rather rare events. Also theorists claim to have given an explanation on astronomical grounds for the recent four ages—but then if there had been five, might they not have discovered a different but equally cogent reason for there having been five? So we cannot be quite sure that there may not be more of them to come within a few tens of thousands of years. However, these things are trivial, for as first Scandinavia, then Scotland and then England became uninhabitable, so the climate further south would improve; rain would fall in the Sahara, agriculture would flourish there and a general shift of populations southwards would leave things much as they are.

In this connection the direct influence that civilized man has had on geography may be noted. Less than ten thousand years ago England was connected with Europe over what is now the North Sea. This region was gradually drowned and, but for the direct action of man, most of Holland, and the English fen country would by now also have been drowned and indistinguishable from the North Sea. But these are comparatively minor matters, for the evidence of the past shows that the sea level has altered up and down quite considerably on account of the varying amount of ice locked up at the poles, and it is evident that as trivial a change as fifty feet in the level of the sea would entirely defeat man's efforts to preserve the low lying regions, or conversely with a rather larger change in the other direction would make it impossible to preserve Britain as an island. Man's direct influence on geography is really quite negligible. On the other hand his indirect influence on geography has been more considerable, since he has made very perceptible changes in climate by the felling of forests. This felling tends to remove the spongy cover of the ground which acts as a reservoir for water, and it leads to a consequent erosion of his fields. All this is now very much on the public conscience, and some remedies are being found, so I shall not go into it. It does, however, illustrate how the short-term increase in the area under cultivation may be very detrimental to agriculture in the long run.

It is more interesting to inquire whether man may hope to gain any direct control over climate. In the first place it can safely be said that it is quite impossible that he should directly cool the tropics and simultaneously warm the northerly regions, for it must always be true that the average temperature will be higher in the lower latitudes. If there were to be another ice age, which would cool the tropics, this could only be at the expense of still further chilling the poles. But there are exceptions to this general principle, which are brought about by the circulation of the ocean. The Gulf Stream has given north-western Europe a climate roughly equivalent to that found in other parts of the world from ten to fifteen degrees further south, and conversely the Humboldt Current of cold water off the west coast of South America has made northern Chile and southern Peru much more habitable than other parts of the tropics. Currents on this vast scale are of course uncontrollable, but there are other cases where control, though it may be impossible, is not unimaginably impossible. For example, the Bering Strait is only fifty miles wide and not very deep, and great currents flow through it to and from the Arctic Sea. If it were blocked, these currents would cease, and it may be that the climate of north-west America and north-east Asia would be considerably changed—though I have no idea whether it would be for the better or for the worse. If it could be confidently calculated that it would be very much for the better, so that an area of the size of a small continent was made habitable through the blocking, it might become worth while considering the devotion of quite a fraction of the whole world's resources to the stupendous task.

On an altogether smaller scale there is the question of rain-making. As is well known, it has recently been found that when there are heavy clouds which are nearly raining, they may be made to rain with the help of solid carbon dioxide powder. The cloud was very near the point of instability, and the small stimulus was enough to topple it over. The most obvious use that could be made of the process is not actually to make rain, but rather to choose the place where it shall fall. Thus unwanted rain might be made to fall in the sea, or it might be possible for a district needing rain to get it out of clouds, which would otherwise have only let their rain fall later in another region; political complications seem very possible if this should be done.

This rain-making depends on the air conditions being on the verge of instability, for then practically no energy is needed to make the clouds rain. It is quite different in ordinary conditions of weather, for then it would call for an enormous expenditure of energy to change the weather, whether to make it rain or to stop it from raining. At the present time we do not know at all how we should go about the business, even if we did have the energy available, but in spite of this ignorance we can still confidently say, from the general principle of energy, that it would not be worth while trying. What would be the use of filling a large water reservoir by means of rain, if it took more than all the hydro-electric power derived from that reservoir to make the rain? A rainy season in the centre of the Sahara, which would be good for agriculture, might for all we know be produced by the use of a million tons of coal, but it would certainly call for far less to irrigate the desert by distilling water at the shores of the Mediterranean and carrying it there by road or through a pipe. In the light of these considerations it does not seem likely that man can ever do a great deal about directly altering his climates.

A thing that will assume enormous importance quite soon is the exhaustion of our fuel resources. Coal and oil have been accumulating in the earth for over five hundred million years, and, at the present rates of demand for mechanical power, the estimates are that oil will be all gone in about a century, and coal probably in a good deal less than five hundred years. For the present purpose it does not matter if these are under-estimates; they could be doubled or trebled and still not affect the argument. Mechanical power comes from our reserves of energy, and we are squandering our energy capital quite recklessly; it will very soon be all gone, and in the long run we shall have to live from year to year on our earnings. All the energy from coal and oil came from the conversion of the energy of sunlight into the chemical energy contained in plants; the conversion is not very efficient, and left to itself the vegetable kingdom certainly will not year by year produce even remotely enough energy to satisfy our present scale of demand. Water power is the only really big present source of energy that can be counted as income and not capital; it derives its energy from sunlight too, through the evaporation of water in the ocean and its precipitation as rain on the mountain tops. Though water power is important, it contributes a fairly small fraction to the present demands of the world, and estimates do not suggest that it could ever expand so as to supply the whole of the demands. During the long run of a million years, a great deal more energy will be needed.

It is worth giving consideration in a little detail to the shortage of energy, both because of its tremendous importance to human life, and because it is possible to speak about it with some confidence. There are going to be many shortages of all sorts of things in the future; for example, metal mines will be exhausted, and many of the metals we now use will run short some day—some of them in the very near future—but it can reasonably be expected that fairly good substitutes will be found for them. But energy is different; there is no substitute for energy, and no way of creating it. It is no use adopting the Micawber attitude that "something will turn up", an attitude which may be admissible over the shortages of metals; but not for energy, because for that nothing *can* turn up. The utmost that can be done is to discover the key to unlock some known but at present unavailable source of energy. This is true even of what many will regard as a newly discovered source, atomic energy; for the existence of this energy has been long known, and the novelty is that the key has only recently been found. In the light of these considerations, I shall devote a little space to considering what are the future prospects of energy for the use of humanity, and from what sources it may be derived.

Atomic energy has been much discussed in recent years as a source of power which may ultimately replace coal. It is certainly too early to estimate this with confidence, but the prospects are really not very bright. The only method of getting atomic power, which is at present in sight, is from uranium. Now uranium is a fairly common element, commoner than silver but not as common as lead, but present estimates suggest that the total energy that could be derived from the earth's uranium is very roughly as much as has come and will come from coal; it is unlikely to be ten times as much, and it is certainly not a thousand times as much, so that it would not help in the long ages to come. Moreover there are very few mines where it is strongly concentrated, and for the rest it would be a costly and destructive business, to work over vast bodies of poor ore in order to win relatively tiny quantities of uranium.

The matter is made only a little better by the existence of the rather commoner element thorium, which has not yet been tamed into giving up atomic energy, though this will probably happen some day. The production of energy from uranium or thorium, as far as we can judge, will always have to be done in "piles", which have to be very large units if they are to work at all, so that the distribution of the power to the users is itself quite a problem. Furthermore there are really formidable secondary difficulties associated with making energy from uranium. There is the familiar political danger that it is impossible to get the power without at the same time making large amounts of explosive material suitable for atom bombs. Then also there are made large quantities of intensely radio-active fission products, the cinders of the furnace, and even at the present time, when developments are still almost rudimentary, the disposal of these cinders is a formidable problem. On the whole then the prospects of power from uranium are not very good; it may be a useful palliative in the energy shortage, but it almost certainly will not provide a long-term solution.

It is well known that there may be a possibility of making atomic explosives from hydrogen, and, since this is a source of energy, it might some day be made into a fuel to yield power. It is the isotope, heavy hydrogen, that would be used, and though its proportion in hydrogen, or in water, is very small, still there are broadly speaking unlimited stocks of it. In practice it takes a good deal of energy to separate it out from the ordinary hydrogen, but the amount is trivial compared to the energy it would yield after the separation. There seems little doubt that the heavy hydrogen will some day be made to explode with the help of a suitable detonator, but this would be useless as a source of power; for that purpose it is necessary that it should be made to "burn" slowly, and this may be an insoluble problem. If, however, it could be done, it might yield a permanent solution of the fuel problem.

To complete the picture of atomic energy, there is ordinary hydrogen, which potentially contains most energy of all. It is ordinary hydrogen atoms that yield the energy which keeps the sun and stars hot; this they

do through a series of rather complicated reactions at enormous temperatures which gradually unite them into helium through the agency of atoms such as carbon and nitrogen. As far as we can judge this energy is permanently locked up in the case of the hydrogen on earth, and perhaps it is a good thing; for if it were not so, there would be quite a probability some day of an explosion which would wreck the whole earth, and indeed the solar system. The "burning" of heavy hydrogen must always be controllable, because it has to be preceded by the laborious separation of it from ordinary hydrogen; but if it were possible easily to "burn" ordinary hydrogen, sooner or later some madman, or perhaps a disappointed would-be world-dictator, would set fire to the sea in such an uncontrollable way that the wave of burning would consume all the hydrogen on the earth. A rough calculation shows that the energy would be enough to make the earth shine for more than ten years as brightly as the sun does now.*
It would make the solar system into a very respectable new star. On the whole it is very satisfactory that we are never likely to be able to "burn" our hydrogen.

Finally, there is a conceivable source of energy in the

* For this kind of calculation it is convenient to use the principle of relativity and count energy by its weight in tons. The sun radiates four million tons of energy every second. To estimate the amount of hydrogen on the earth, I assume that the sea holds most of it. The amount of water in the sea has been estimated at about 1.4×10^{18} tons. One-ninth of this weight is hydrogen, and of each atom of hydrogen eight parts in a thousand are available for atomic energy. The result is 1.2×10^{15} tons of energy, or as much as is given by the sun in three hundred million seconds, which is about ten years.

annihilation of matter. This would give a supply hundreds of times more potent than the "burning" of hydrogen into helium, and it would presumably be hundreds of times more devastating, but it is quite unknown whether it can happen at all, even in the hot interiors of stars. It is safe to say that long before this source could be used, some of the milder forms of atomic energy would either have been made available, or else would have destroyed the world.

Since the prospect of getting atomic power on a really large scale seems not very good, and since water power, which is much the most straightforward source of energy, is going to be inadequate, it is important to consider what other sources might be exploited. Possible sources, in addition to vegetation, are the direct use of sunlight, wind, tides, the interior heat of the earth, and the cold water at the bottom of the sea. Some of these can never provide large powers, and others suffer from being very diffuse in their distribution, but they all deserve consideration.

The internal heat of the earth is already being exploited at an installation in Italy, where steam is raised by pumping water into hot fissures in the earth. There may be other places where this could be done, though it is hardly likely to be on a large scale. Indeed in principle it would be possible to use any volcano as the furnace of a power station, but it is hardly a practical proposition in view of the unreliable habits of volcanoes. The existence of volcanoes is attributed to deep cracks in the ground, which at irregular intervals of time let in water to a depth, where it is boiled under pressure so that it explodes out again. This suggests the possibility that man might directly tap this source of heat by, so to speak, making artificial volcanoes which he controlled so that they never reached the surface. Heat does not flow out from the centre of the earth very fast, and nothing that man might do could affect this rate, since he can only hope to work on the outermost few miles of the earth's shell. Estimated on a world scale the total energy available is not very great, and the best he could hope for would be to make a few deep borings and raise steam in them. He might hope to keep these borings under control, but even if he was successful in this, there would still be a price to pay, for his disturbance of the temperatures in the earth's shell would almost certainly sooner or later lead to earthquakes. In the light of these considerations not much can be expected from the earth's internal heat, beyond a few more stations like the one in Italy.

An experimental installation has been set up, or at any rate proposed, with the aim of deriving power from the difference in temperature between the surface water of the sea and the water at the bottom. Wherever there is any temperature difference it is theoretically possible to get power from it, but the amount depends on the magnitude of this difference. In the depths of the oceans in all latitudes the water is only a degree or two above freezing, and in the tropics at the surface it is perhaps 80° F., so that there is no great margin to work on, and enormous quantities of water would have to be handled to get any reasonable amount of power. The possibility

THE NEXT MILLION YEARS

of this power is guaranteed by basic theory, but I do not know what mechanism would actually realize it. It would only be feasible to tap this source in special places, such as tropical oceanic islands, for only in such places would there be a high surface temperature together with proximity to the cold ocean depths. It may be conjectured that this source of energy would be too expensive to be much used.

The wind blows on account of unequal heating of different parts of the earth, so that its energy is derived from sunlight, like that of water-power or of the fuels we are now burning up. It could provide considerable amounts of power by means of windmills. The difficulty is that each windmill can only collect a rather small amount of energy, and this at irregular times, and so, to make the wind a really useful source of power, some method of storing the energy is quite essential. The most straightforward way of doing this is to have a large number of mills, which pump water up into a reservoir whenever the wind blows, and then when the power is demanded, the water of this reservoir is used hydro-electrically. This tends to limit such a scheme to hilly country, where the differences in level make it possible to construct the reservoirs. A more profitable development, not yet in sight, would be the invention of some really cheap way of storing energy chemically; the ordinary electric storage battery is exactly the kind of thing needed, but it is far too expensive. Indeed if any cheap device should be discovered, whether depending on mechanical or electrical or chemical or any other principles, which would store large quantities of energy reasonably efficiently, it would go a very long way towards solving the whole power problem, whether the energy came from the wind or any other source. Assuming such an invention made and applied to the wind, the economic picture of the world would be very different from what it is now, because wealth will tend to be associated with easy power supplies. Since it is always likely to be wasteful to transmit power over long distances, it would be the windy regions of the earth that would flourish; these would include the areas of the trade winds, many deserts where a wind springs up every day, and the stormy areas which are found in high northern and southern latitudes. In these regions there would be set up great rows and ranks of windmills together with the devices for storing the energy. How far schemes of this kind may develop will depend on how successfully the storage problem is solved, and it is probable that if it is solved, the wind will be an important contributor to the world's power problem.

The tides are an obvious possible source of power. An interesting point is that they would tap a source of energy quite different from the other possibilities, for their power would be partly derived from the rotational energy of the earth and partly from the orbital energy of the moon; the use of tidal energy slightly lengthens both the day and the month. Some use is of course already made of the tides for power; for example, the English fens are pumped out by opening the sluice gates at low tide and shutting them at high tide. There are also quite a number of proposals, where the terrain is suitable as on the River Severn, for making barrages which trap the tide at high water and generate hydro-electric power from it. Even under the most favourable circumstances there is the inconvenience of the fortnightly cycle, for during each fortnight the tides vary in height by a factor of three between springs and neaps, and high tide is at varying times of day, so that some form of energy storage is quite essential. This applies even when the conditions are most favourable, and the difficulty would be far greater if it were attempted to collect power from the tides on the open coasts. The same difficulty would arise there that arises with wind power, of having a very large amount of energy spread very diffusely, so that a really cheap method of storage is essential if its collection is ever to be at all practical. It must also be remembered that the tides in the open ocean are only a foot or two in height, and that there are not very many parts of the earth where the coastal configuration enhances them to a magnitude that would be easy to exploit for power. Britain is one of these, and if once the storage problem can be solved, here and in the other favoured areas the tides could make a useful contribution to the power problem.

The direct use of sunlight would be one of the most effective ways of getting power. Already in suitable climates it is used for heating water by absorbing heat on the blackened surfaces of water tanks, but this is trivial compared to what might be hoped for. One obvious way of getting power would be to use the heat of the sun to raise steam, by concentrating it on to the surface of a boiler by means of a burning glass, or more probably a reflector. It would be a formidable problem, for the total amount of heat falling on a square yard facing the sun is about enough in each minute of time to evaporate only a quarter of an ounce of boiling water into steam, so that to make an engine of reasonable size a very large area would be needed. On the other hand, the efficiency of the engine would be very good, for the temperature of sunlight is six thousand degrees centigrade; this is the actual temperature of the sun's surface, and it signifies that the sun's rays could ideally raise a boiler to this temperature. It is this temperature that matters for efficiency, and it means that the temperature of the boiler need only be limited by the strength of the materials it can be made of. The result might be to convert perhaps a fifth of the heat into power. On this basis something like a third of a horse power could ideally be obtained from a square yard; this is not ground area but area measured facing directly towards the sun, which of course will demand a bigger area on the ground.

The possibility of getting the solar energy directly in this way is impressive, but the technical difficulties would be very formidable indeed. Apart from all the ordinary difficulties of big engineering projects, the chief would probably be that it would be necessary to concentrate the heat from a good many square yards on to a rather small focus, for it is only so that the heat losses could be avoided which would destroy the engine's efficiency; and this must somehow be done in spite of the sun's motion all through the day, and its different height at different times of year. Deserts, where the sun always shines and there is no rainy season, would be the best places for solar engines. The power would only come in daytime, and this would be less inconvenient than tidal power, but still it would be nearly essential to be able to store it. Altogether it would be a tremendous undertaking. There may of course be discovered other ways of getting energy out of sunlight; for example, there may be chemical processes, which would imitate those of the vegetables, but more efficiently. There is also the possibility of getting the energy photo-electrically, that is to say by causing the light directly to make electric currents. At present this is a hopelessly inefficient method, but it cannot be excluded that some new idea might make it feasible, and then it would probably be the best of all. Of all the possible ways of collecting energy, the direct use of sunlight is the most promising.

Finally, it may well prove that the various devices discussed above are all of them too complicated and troublesome to be really practical, and that it is best to exploit the method used by nature, the vegetable. There would have to be vast plantations, producing potatoes or some such plants in enormous quantities, which could be made into industrial alcohol for power. Or perhaps it might be possible to exploit the ocean, by collecting the microscopic vegetables floating on its surface. The quantity of this plankton must be vast, but it is spread very thin, and the collection would be a very difficult problem. However that may be, and wherever the vegetables grew, there would be all the trouble over bad seasons and pestilences that we know too well already, and it is possible that enormous greenhouses, in which the plants grew under accurately controlled conditions might pay better; but whatever was found best would have to be on a vast scale, because of the comparative inefficiency of the vegetable in converting sunlight into energy.

I think this completes the list of all reasonably possible sources of energy, and apart from ordinary water-power the results are not encouraging. This is hardly surprising, for it certainly involves a great deal more work to live on income than on the accumulated capital of geological ages. Our present civilization is largely based on the provision of mechanical power, and if it is to continue, it would seem likely that a good fraction of humanity will have to be engaged in collecting energy, either by minding vast numbers of machines, or by tending vegetables in plantations; it will have to be a far greater number than those now engaged in mines and power-stations. It is rather likely that the natural inefficiency of mankind will prevent him from realizing to the full the possibilities of winning energy out of nature, and that he will often find that he has to get on with much less of it.

Turning now to other questions of the future conditions of the world, it is of course likely that many technical inventions, both of utility and of luxury, will be made, which may profoundly alter the detail of human life. As I have already explained in an earlier chapter, however, and as will be developed more fully later, these are only to be regarded as details superposed on the immensely more important questions of population and of human nature. It is therefore not worth entering on wild speculations about them, for such speculations would surely be as wrong as the speculations of a natural philosopher of two centuries ago would have been about our present conditions. It may be noticed, however, that the biological sciences, which in the nineteenth century rather lagged behind the physical, are beginning to show promise of quite astonishing new advances.

The proper consideration of these biological advances must for the most part be deferred to later chapters, but I may list some of them here without discussion. There is first the possibility of new sources of food; for example if grass or wood could be rendered edible, it is safe to say that there would be immediately a great increase in the population of the world. Then there is the probability that medical science will continue still further the great triumphs it can already claim in the conquest of disease. It is also not impossible that medical science might succeed in materially lengthening life without senility, though in a world of overcrowded population it is not very clear what would be gained. Looking a little deeper there is the possibility of substantially altering the intellectual and moral natures of individuals by some sort of hormonal injections; already great effects have been produced on animals. Finally, as the most curious speculation of all, it is not quite impossible that it may one day be feasible to select in advance the sex of each child that is to be born. Whether the decision is made by the parents, or by their rulers, this suggests the probability of a great unbalance in the populations of the world. Before discussing these matters, however, it is necessary to look deeper into man's nature, and this will be the subject of the next three chapters.

V

THE SPECIES *HOMO SAPIENS*

THE central thing that goes to make up history is not the external conditions of the world, but the nature of man himself, and this will be the subject of the present and the next two chapters. Man is an animal, but a social animal, and in discussing him it is convenient to draw a line between his own inherent nature, and the way he is influenced by the society in which he lives; the distinction between these is of course only a very rough one and there is much overlapping. In the present chapter, I shall regard man as a biological specimen like any wild animal, and in the next I shall consider how he is influenced by the society round him; but then the important question arises as to the balance between these rival influences, and that subject is reserved for the chapter following.

One of the interesting recent developments of geology has been the possibility of dating the past with far greater accuracy than could be done before. Several different methods have been applied, which agree in their broad results, but I shall not go into them here. The dating has been specially accurate in relation to the recent ice ages and this means that it can be applied with some confidence to existing animals and plants. One of the things discovered from the study of the remains of this period is the answer to the question—in the evolution of life, how long does it take to make a new species? The answer is a million years. That is the reason for the title I have chosen for this essay—for a million years to come we have got to put up with all the defects in man's nature as it is now.

It is hardly necessary to say that there is nothing very exact about this million years. Some species change more quickly and some more slowly, but it does seem to be a good rough rule, and curiously enough it seems to apply more or less irrespective of the number of generations in the million years, which of course would be immensely more for an insect or a rat than for a buffalo or a man. The million years may perhaps not be a very close estimate; it might even prove to be only half a million or possibly two million for man, but it is hardly possible it should be as short as a hundred thousand years. So it is good enough to assume that it is a million years, and if this is an over-estimate, the reduced length of time is still long enough to give a fair average of human history.

It is a vexed question exactly what the word *species* means and many answers have been given, but perhaps the best answer so far was that given by the cynic who stated "A species is what a trained taxonomist says is a species". This does not seem to advance the subject very much, but it is a fact that the trained taxonomists, who have frequently disagreed among themselves about other species, are all agreed that the species *homo sapiens* includes all the races of humanity. There are, however, obvious differences of complexion and feature among

78

them, which constitute them as *varieties* of the species. Since in the history of evolution a variety is the starting point for the formation of a new species, it might be imagined that, if one of the several races were completely isolated from the rest, it would slowly turn into a separate species; but there is no chance of any such isolation, and anyhow if it should occur it would take a million years to make the species, and so it would fall beyond the span of time I am considering.

It is natural to believe that, when there are such obvious differences of complexion between the various races of man, there should go with these some differences in brain and in mental characters, but the psychologists and anthropologists have found it difficult to detect them. In mental characters the range of variation inside each individual race is very wide indeed, so much so that it entirely submerges any difference in racial characters, if such there should be. So it is not useful to give any consideration to differences of race; in every race there are highly intelligent people and very stupid ones, and all mankind display the same characteristics of pugnacity, ambition, envy, laziness, selfishness, unselfishness, loyalty, kindliness, sociability, sense of humour and so on. There are of course obvious differences in behaviour between individuals on account of differences of condition and of training or education, which I shall discuss in the next chapter, but it is correct to say that man really is one species and that as such it will take a million years before anything notably different will arise in his nature. This is a fixed point, which can be taken as the central thing that makes possible the prediction of his history for a million years, and no longer.

As an animal, man is subject to all the rules of heredity, the general principles of which have been fully worked out, even though much of the detail is still unknown for the human species. Of course it has always been obvious that there is a natural tendency for offspring to inherit qualities from their parents, but the principles, discovered about a hundred years ago by Mendel though not widely known till the present century, have defined the situation very much more precisely. This is not the place to go deeply into the subject, and I shall only cite a few points which are germane to my purpose.

The central feature of the Mendelian theory is the *gene*, which is the unit of heredity. A gene is usually only recognized by its effect on the bodily development of the animal, but the chromosomes in the animal's cells, which are strings of genes, are easily visible under the microscope, and in some cases the position of a gene on its chromosome is fairly well known; so it may be said that the genes are particles of living matter which are very nearly visible. The germ cell of every animal contains a very large number of genes, and these dictate all the details of the animal's development, such as whether it is to be tall or short, light or dark, and so on. There are known rules, some of them quite complicated, but still perfectly definite, which determine how the genes are handed on from one generation to the next. The new generation has to have a complete outfit of genes, and this it accomplishes by drawing each of them from its father or its mother, but not from both; it is pure chance which parent contributes any particular gene. So the offspring contains a mixture of the genes of its parents, and therefore develops a mixture of their qualities. The genes of man, like those of every other animal, control the development of every part of his body, and this includes his brain, and since the quality of the brain determines all the natural mental characteristics, these also fall under the control of the Mendelian laws of inheritance. There is no doubt of this, but it must be confessed that up to the present time little is known about the detail of the actual genes of humanity.

It is through the inheritance of qualities useful in the struggle for life that natural selection works, but, with the old vague ideas about heredity, it was rather hard to see how a race of animals could be really benefited by any valuable character that might appear in one of its members. This animal's mate would not usually have the character, and so, according to the old ideas, the off-spring would be expected to have it to half the extent of the favoured parent, the second generation to only a quarter, and so on. It seemed therefore that the character would be rapidly diluted in the succeeding generations, and it was hard to see how, in the long run, it could re-tain enough value to give any significant advantage. This difficulty is cleared up by the Mendelian law. The parent with the valuable character has a gene for that character, which it transmits on the average to only half its offspring, but those that do receive it receive it to the full; the rest do not get it at all. Thus, for those who get it, there is no dilution in the quality; it continues at full

F   81

strength and is able to give to its possessors the full advantage that it confers in the struggle for life. There is thus a good prospect that the valuable quality may, so to speak, become anchored to the species by being incorporated among the genes of the majority of its members. If a dictator should ever aspire to bring about some really permanent change in humanity, he could do it if, and only if, he knew how to alter some of the human genes, for only so could the changed quality become anchored as a fixed character of the race.

Genes retain their constant character for many generations of the animal cells, but they do occasionally change, and it is by the cumulation of these chance *mutations* that a new species may arise. Recently it has been found possible to increase very much the frequency with which mutations occur, so that one might aspire to make much more rapid changes in the characteristics of an animal species than those which occur in nature. The method is to expose the germ cells to X-rays, or to certain chemicals, which disturb the process of cell division, so that the new cells may possess one or more changed genes. The process is in no way controlled by the experimenter; the X-rays simply stir things up, so that an arbitrary change results, which he can then study and exploit. Quite a number of mutations have been produced in insects by these means, but most of them have been deleterious. This is not surprising because an animal is a very delicately balanced mechanism, with its constitution continually kept up to the mark by the stringent conditions of life, and a large change in any part of its structure is far more likely to upset the balance than to improve it. Similar changes could no doubt be induced by X-rays in the human genes, but there too it is far more likely that the consequence would be deleterious than beneficial, because of the upset of the balance of human qualities. To make any large beneficial change in one step, it would be necessary to make favourable changes simultaneously in several genes, and there is practically no chance that any X-ray dose could do this, without at the same time damaging some of the other genes in the human germ cell. Even if we knew what we wanted, the prospects of improving human nature in a single step, or even in several steps, by artificial means, are so small that they can be left right out of account. The only prospect of improvement must be by taking advantage of the rare occasions when a *small* beneficial mutation happens to arise.

Even without mutation there is a tendency for animals to degenerate, and this in spite of the constancy of the influence which each gene exerts in the formation of the animal's body. The reason is that in many cases several genes have to co-operate together for the correct formation of one of its organs. The classic example is the eye of the fruit-fly. This develops under the simultaneous control of many separate genes, and in consequence it can exhibit a great variety of defects, each attributable to the lack of one or more of them. In the laboratory these defects can be preserved and studied, but in wild life natural selection is continually destroying the insects with bad eyes, and thus the species is being kept up to the mark. This example suggests a speculation about the human eye, though of course that is an entirely different and much more wonderful organ than the eye of any insect. Man's life depends very much on his eyesight, and in the long past anyone with defective sight would have had a distinctly lowered chance of survival. Nature must have been continually keeping the human eye up to the mark in this way. But fifteen generations ago spectacles were invented, and at once some eye defects, such as shortsightedness, ceased to be a serious handicap. There is thus now no check against shortsightedness, and it is a fairly safe forecast that in another hundred generations or two, this defect will be even commoner than it is now. This speculation illustrates how any human quality may be expected to degenerate, unless it is being disciplined all the time by the stringent test of natural selection.

There is one other principle in the laws of heredity, which calls for special comment here, since, though it is familiar to biologists, it is often unrecognized by laymen. It is known as the principle of the *Non-inheritance of Acquired Characters*. This signifies that any change acquired by an animal during the course of its life is never passed on to its offspring; the simplest example is a mutilation, but the same rule applies to a skill of any kind that the animal learns during the course of its life. The subject has been hotly debated among biologists during the past seventy years, and all authoritative opinion is now agreed that such effects are not inherited. However, it is almost impossible ever to prove a negative, and at intervals new examples are still cited, which are claimed as showing that characters acquired by an animal during adult life have been handed on to the offspring. Most of them do not stand the test of close examination, but even if there should be a residue of valid examples—and there is no reason to believe that there is—it is safe to say that a phenomenon, so difficult to prove and so rare in its occurrence, cannot have played any important part in the development of life on earth. The non-inheritance of acquired characters is just what would be expected from the Mendelian theory. The new generation derives its genes from those of its parents, and these parental genes were laid down before the parents were born; and they will not in any way have been affected by his and her later experiences, including those experiences which occurred before the procreation of the offspring.

It may appear surprising that it took so long to establish definitely such a simple principle as this of the non-inheritance of acquired characters, but closer consideration shows that the matter is rather more subtle than might appear at first sight. Many genes govern quite straightforward qualities, such as the colour of a man's eyes, or the fact that after twenty years he is to start growing a red beard, but there are others which are by no means so obvious, such as those that determine tendencies of character; these tendencies may never exhibit their effects at all, unless an appropriate external situation should arise to evoke them.

This may be illustrated by an example. Dogs vary very much in the ease with which they learn tricks, but the dog with this gift will never show it unless he is taught the tricks. The tricks themselves are obviously acquired characters, but the ease of learning them may be innate. A naïve observer might think that the puppy learnt the tricks easily, because its parent had been taught many tricks, whereas actually all that has happened is that the parent did the tricks because it had the innate capacity for learning them easily, and it has handed on this innate capacity to its offspring. Notice the contrast in the action of the dog's trainer according to whether acquired characters are inherited or not. Suppose that there are two puppies of the same litter, and that one of them learns tricks much more easily than the other. The second would need far more training than the first, and if the trainer really believed that acquired characters were inherited, he would expect that the descendants of this second dog would reflect the consequence of this greater training by themselves being easier to train; actually of course the exact opposite will be true, since the descendants of the dog which needed little training will learn the tricks more easily. This example illustrates a point of immense importance to humanity that is all too often overlooked. I shall return to it in the next chapter.

The Mendelian laws are perfectly precise, and I have been describing them almost as though they had the same sort of certainty as the fundamental laws of physics, and I have not yet taken much notice of the element of chance that enters into them very intimately. This element of course arises from the way in which the offspring derives half its genes at random from one parent and half from the other. To those who are not very familiar with the principles of probability it may appear that this fact will spoil the force of the whole argument, since chance means uncertainty, and therefore it might appear to destroy the validity of any law. Furthermore the matter goes very much deeper, for pure chance plays a great part in all the subsequent life of every animal, as well as in its procreation, so it might well be asked what is the use of being so definite about the laws of heredity when they are always going to be dominated by the fortuitous circumstances of the animal's life. Such reasoning overlooks the influence of large numbers, which will reduce the most wildly various individual experiences to a nearly steady average. Since there may be readers who are not familiar with the extraordinary cogency of arguments based on probability, I shall devote a little space to the subject, and in the course of it a number of other points of interest will emerge.

The fortuitous occurrences in the events of life are what mostly attract our attention, but in the long run it is the law of large numbers that counts, the law that the result of a large number of chance events tends to approach towards an average. For example, the player at the roulette table remembers chiefly the occasions on which he made large gains or losses, and he is apt to forget that in a lifetime of gambling the actual result will almost certainly be that he has lost a small percentage of his total stakes, the percentage levied because the roulette board has a "zero" number which biases it slightly in favour of the casino manager. In the long run it is this bias that counts, and the life of the human race on earth is certainly to be counted as a very long run, so that the bias is to be reckoned as the really important thing in it, and not the chance good or ill fortunes of individuals, or even of nations and races.

At the risk of over-elaborating the matter, I will take another example from games of chance which brings out some interesting points. Five men sit down to a game of pure chance, each contributing the same capital sum, and they play according to the rule that anyone retires when he has lost all his capital. It is then a certainty that in the end one of the five will win the capital of all the others; it is of course pure chance which of them it will be. If the stakes allowed on each hand are a large fraction of the capital the game will be short, while if they are restricted to being small, the game may take a long time, but the final result will be the same. Now suppose that the capital of the players is no longer required to be equal; at once there is a bias in favour of the rich man. In the course of the game each player has ups and downs, but the rich man may sometimes recover from a loss that would have bankrupted the poor man, so that he is definitely more likely to be the winner. The moral of this is that, if there are two nations in which the individuals of both have equal merits in the struggle for life, the one with the larger population will tend to have the advantage over the other. In this I am not considering the fact that the larger nation might enlist more battalions so as to conquer the other; it is just that through its greater population it can stand great misfortunes and still come back from them to prosperity, whereas those misfortunes would have totally destroyed the smaller nation.

Returning to the game of the five players, suppose

once again that they have equal capital but that, though it is still mainly a game of chance, there is a small measure of skill in it. It now becomes likely, though not certain, that it will be the most skilful player among the five who will win. The skill may be of any kind; it may be that, through greater intelligence, he can better estimate the chances so as appropriately to vary his stake at any stage of the game, or it may be that he possesses fingers which make undetected cheating possible—I am considering simply who will win the game, not the question of the winner's moral character. If the stakes are high, so that the game is likely to be short, there may not be enough time for his skill to tell, and then he will have little advantage against the operations of pure chance, but if the game takes a long time it becomes exceedingly probable that it will be the skilful player who will win. The moral of this is that natural advantages in the struggle for life will tell in the long run. For the individual animal or man the stakes are often too high, so that he may be killed before his merit can show itself, but the history of the world is a long thing, and it is concerned not with individuals but with large numbers of individuals. Because of its immense scale the game of life is to be regarded as a game of small stakes, so that it becomes very nearly a certainty that the inherent qualities of the race will be what counts, and not the accidents to which individuals or even nations are exposed.

Before passing on there is one further point to be made. I have explained how the operation of chance becomes comparatively unimportant for large numbers, and it is very pertinent to ask how large the numbers must be. If, for instance, it was only true for millions of millions, it might be felt that in any reasonable span of time chance would still be dominant. Such a very general question can only be given a vague answer, but the answer is, that the number usually need not be at all large for the chances to average out. With the typical example of spinning a coin, even a quite small number like ten will almost count as a large number in the sense that, if the coin is spun ten times, the number of heads will rarely be more than two away from five, which is the average number of heads. In most matters concerned with probability three or four count as small numbers, ten as a fairly large number and a hundred as a very large number. There are of course exceptions to this statement—as, for example, when the chance in favour of some very rare occurrence is being considered —but it will do in giving a general picture of the subject.

In the light of all these considerations how does man stand in the animal kingdom in regard to the heritable qualities which are to help him to survive? Physically he is a poor thing, neither so strong, nor so swift, nor so tough as other animals, nor with effective means of defence. On the physical side the only claims he can make to any superiority in the animal kingdom are his eyes which have a refinement of perception above that of all mammals and probably of most birds, his hands which are one of the most versatile and delicate tools in nature, and the gift of speech, which has such tremendous social importance. But these things are relatively

trivial, since the essential point of man, the new phenomenon in nature, is his intellect, associated with his enormously exaggerated brain. It is therefore in relation to his
intellect that all man's other qualities must be considered.

In the essential matter of survival there are two things
needed, the survival of the individual and the survival
of the race. We are all very well endowed with deep
instincts for both, and curiously enough we are ashamed
of both these instincts. As to the survival of the individual we have a very strong, intimate and deep fear of
death, evoked by any form of danger; it is not a thing
we boast about, but it is certainly a very essential quality
for survival, and as such it is to be regarded as important
and valuable. For the reproduction of the race, there are
two instincts needed, the sexual and the parental, and the
way these are organized is to say the least curious. The
sexual instinct, though much complicated by all sorts of
taboos, is for most of mankind nearly as violent as the
fear of death, though it has the advantage of being
pleasant instead of unpleasant. Among animals it brings
about the inevitable consequence of reproduction, and
until very recently the same was true for man, so that
the Malthusian increase of population was assured. This
is still true for a large proportion of the human race, but
the existence of birth-control has entirely altered the
situation among the more highly developed peoples.
The consequence has been to make reproduction depend
for them not on an intense instinctive impulse, but rather
on intellectual reasoning, and this for very many people
is an exceedingly tepid motive.

The parental instinct is also somewhat ineffective, because for the majority it is only strongly stimulated by the presence of the children; that is to say, it is very important in preserving them, but it does not make any such clamant call on the emotions to beget them. It has not the same intensive compulsion as the sexual instinct, and this is not very surprising because of its very different function. No one can feel any very intense emotion continuously for more than a short time; whether it is pleasure or pain, anger or grief or fear, the sharp edge of it fades in a few days, whereas the parental instinct has got to work effectively for fifteen or twenty years, if it is to serve the survival of the race. It is therefore hardly surprising that it should be steady and continuous, but not so intense an instinct as the sexual instinct or as the fear of death.

It is a matter of great importance that the procreative instincts are at present ineffective in maintaining our population, but I shall defer considering this till a later chapter after some of the other attributes of humanity have been discussed. Besides the instincts that I have dealt with, there are of course many other qualities, some of them not so instinctive, which are important for survival. One may be selected as pre-eminent. Man is superior to all other animals in his readiness to try experiments. Many of the higher animals, like him, can learn by experience, but if they are placed in an unfamiliar situation they are lost, whereas a man will always try to think it out and will often find a solution of his difficulties. This flexibility of mind, and the adaptability to unforeseen conditions are the main reasons why he has succeeded in dominating the world. They are possessed to very varying degrees by different individuals, but they are clearly of supreme importance in the struggle for life in an ever-changing world.

There are many other qualities, which help survival—and I shall be content to mention only a few of them, some estimable and some the reverse. We value intelligence, honesty, capacity for leadership, and other similar qualities, and we mark our approval by selecting their possessors for promotion. A man is promoted on account of his individual merits, without any thought about the consequences for the distant future. In a less abnormal world than the present, his increased prosperity should lead to the man's having a larger family than those of the less prosperous, so that the good qualities inherited from him should gradually become diffused throughout the population in later generations. At the present time the exact opposite happens all too often, in that he is likely to have a smaller family than the average; in fact success in life is at present antagonistic to success in survival. I shall not consider this further now, since it is to be discussed later.

It is always necessary to remember that nature itself is quite non-moral, and that there are many qualities which we by no means admire, which nevertheless are often regrettably effective in the struggle for life. All through the animal kingdom one of the most successful roles is that of the parasite, and there are states of human society where such a parasite as the professional beggar is as successful as anyone else. Something of the kind is unfortunately true in Britain just now. The people we are really encouraging are not those that we think we are, for a great many of the people who get good promotion are contributing less than their share to the next generation. At present the most efficient way for a man to survive in Britain is to be almost half-witted, completely irresponsible and spending a lot of time in prison, where his health is far better looked after than outside; on coming out with restored health he is ready to beget many further children quite promiscuously, and these "problem children" are then beautifully cared for by the various charitable societies and agencies, until such time as they have grown old enough to carry on the good work for themselves. It is this parasitic type that is at present most favoured in our country; if nothing is done, a point will come where the parasite will kill its host by exhaustion and then of course itself perish miserably and contemptibly through having no one to support it. Now though there may be occasions in human history when something of this kind can happen, there is no fear that it should happen to mankind as a whole; for a parasite is essentially subordinate to some host, and man claims, and claims reasonably, to be master of the world, so that there is nothing for him to be subordinate to. In the long run there is no danger of mankind adopting the role of the parasite.

There is another role, which is not by any means wholly admirable, that may well be specially successful in the struggle for life. This is the role of the hero, using the term not in the modern sense of a man embodying all the virtues, but in the original sense used by Homer. The Homeric hero, who has his counterparts in many other semi-barbaric conditions of life, is brave and reckless, but selfish, undisciplined and something of a bully. He is by taste a leader, but his leadership is often marred by impatience and lack of persistence, so that he fails to carry through to the end any projects which would take a long time. He cares little for the sufferings of others, unless they are his henchmen, whom he looks after out of self-interest. From the present point of view one of his most important characteristics is that he is usually by no means monogamous, but very much the reverse, so that his qualities are likely to be reproduced and multiplied many times in the next generation. Is it possible that in the long run the earth should be wholly peopled by heroes? It is irrelevant that it would be an extraordinarily disagreeable world to live in, for there is nothing in nature to dictate that the world has got to be agreeable. It does, however, seem unlikely that the heroic type can ever become a large fraction of the earth's population, because their qualities do not fit into a society of dense population—rather they prevent its existence—whereas the densely populated countries will dominate the earth by sheer force of numbers. Still there is certainly a place in the world for the heroic type, in particular for their capacity for leadership, and if this quality could be dissociated from some of the other less desirable characteristics of the hero, there would be a great place for him in society. It is not impossible that such a separation of qualities may come about automatically, because the hero who can adapt himself to the civilized life of dense populations will have a better prospect of survival than the one who can only live in a state of semi-barbarism.

This chapter has been mainly devoted to considering the qualities of *homo sapiens* regarded as a species of animal, and species of animals stay roughly fixed in type for something like a million years. In fact, of course, they are slightly changing all the time, but it takes this period before they are different enough to be given a new specific name. It is fitting to end the chapter by speculating on what man's characteristics will be when, at the end of that period, he is sufficiently altered to be regarded as a new species. Will he have become *homo sapientior* or what?

The development of animal species has been studied by means of the fossils of past geological ages, which usually show changes in the animal's form progressing rather steadily through the ages, though the record is often complicated by side branches, which split off into new species or else flourish for a time and then come to nothing. Of course these apparent steady rates of change have not really affected all the members of the species simultaneously. What has happened is that a few of the animals have chanced to develop a small superiority in some way, say an extra turn of speed, and through this advantage they have left more offspring than have the slower animals. The qualities of the members of a species are always spread over a certain range, and the average of the species is being continually dragged upwards, not through an equal change in all the members, but through unequal reproductivity at the two ends of the range. When *homo sapiens* is changing, it will not be by the whole race gaining simultaneously whatever qualities better fit it for survival, but rather by certain types of mankind proving superior to the rest in survival value, so that they contribute a larger proportion to the later generations, and in so doing drag the average qualities of humanity in the same direction.

The first question to consider is whether the varieties of mankind, the white, yellow, brown and black races may not branch off into different species. In the animal kingdom such branching has sometimes arisen through long periods of geographical isolation, and another way has been through the development of infertility in the sexual unions between the most unlike members of the species. Neither of these causes will operate in the case of man, since the whole earth is his habitat, and all the races of mankind are fully fertile together. Climate has been another fruitful cause of the splitting of a species into two. Undoubtedly some of mankind tolerate better a cold climate, and some a hot. Will there then ultimately be a *homo arcticus* and a *homo equatorialis*? It is on the whole unlikely, because man has learnt so well to control his own climates. Even now arctic man can live comfortably in the tropics by cooling his dwellings with the help of physical science, and by resisting the tropical diseases with the help of medical science, and the converse is equally true of equatorial man in the cold regions. It seems likely then that man will not diverge from one species into two on account of climate; indeed a convergence is more probable. Thus although the fair-skinned man can make a success of living in the tropics by taking the trouble to control his climate, his fair skin is still some handicap to him, since it cannot so easily protect him against the direct rays of the sun as would a dark skin. On the other hand, there is not much evidence that a dark skin is a handicap in cold climates. If this contrast of effects is correct, it suggests that in the end man's complexion will be rather dark all over the world.

The physical characteristics of man may of course change with the lapse of time, but it is not likely that they should do so to a great extent, since it is not primarily these qualities that preserve humanity in the struggle for life; even congenital good health and resistance to disease have been largely discounted by medical science. It is the intellectual qualities of man that really matter, and so it will be these that are the most liable to change. In all such qualities there will no doubt be a very wide range of variation between individuals just as there is now, some being clever and some stupid, some good and some bad, and the changes will come about by the increase of the numbers at one end of the range at the expense of those at the other. It is very much a matter of conjecture what those changes will be and I will only mention a few possibilities. General intelligence should always be of value, particularly the unspecialized intelligence that is adaptable to many varieties of purpose; so with some confidence it may be expected that man will become cleverer than he is now. It is by no means so clear that he will become morally better as well, since in a highly competitive world, the sinner has many advantages over the saint. That is disappointing, but it must be remembered that moral codes have differed a good deal at different periods of history, and it is to be expected that future generations will succeed in constructing a moral code, satisfying to the good man of those days and reasonably close to being within attainment by all, however much it may disagree with our own standards.

Another more specialized change may be suggested as probable. Civilization has taught man how to live in dense crowds, and by that very fact those crowds are likely ultimately to constitute a majority of the world's population. Already there are many who prefer this crowded life, but there are others who do not, and these will be gradually eliminated. Life in the crowded condition of cities has many unattractive features, but in the long run these may be overcome, not so much by altering them, but simply by changing the human race into liking them.

Finally I will refer to another quality, which I shall discuss more fully in a later chapter. As has been already pointed out, man's present procreative instincts are failing to reproduce the species in sufficient numbers in many of the more civilized nations. Any of mankind that overcome this failing will increase at the expense of the rest, and there is already a germ for natural selection to work on in the spontaneous wish of some people for children. Whether this reinforcement of the procreative instinct should be dignified by a new specific name I do not know, but, long before the end of a million years, it is almost certain that *homo sapiens* will have changed into *homo paediphilus*.

VI

CREEDS

IN philosophical considerations of the nature of life, a question that has been much debated has been the rivalry between "nature and nurture", that is the question whether the inherent qualities of an animal or the external conditions of its life are more important to it. In fact the argument leads nowhere, because, when two things are both absolutely essential, it simply has no meaning to ask which is the more important. There is not even a clear-cut separation between them, but a separation can be made, if it is accepted that it is only a rough assignment of emphasis, and not a definite splitting of the subject into two. In the previous chapter the emphasis was on man's nature, and it is now the turn for his nurture, in particular for the way human history may be affected by what a man learns from his fellows.

Man shares with the higher animals the capacity for learning, though to an immensely superior degree. The question of how animals learn is much simpler; it has been objectively studied in various ways, in particular by the method of "conditioned reflexes", that is to say, by the study of how by practice (which must always be associated with rewards for success) an animal acquires skill in the performance of tasks which have been set to it. In the experiments the tasks have often been quite unlike anything that would happen in wild life, but the results show the general way in which an animal does learn skills, and the same processes, applied to the conditions of the wild state, undoubtedly assist the animal to survive in the struggle for life. The study of conditioned reflexes has brought out clearly the fact that there is a great variability among different individual animals, in that some learn skills much more easily than others. This is presumably an inherent quality of the individual, but learning is not always a matter of acquiring skills by individual effort, for it often implies definite teaching. This holds particularly in the case of man, but among animals also teaching plays some part. For example, the catching of mice is one of the important things a kitten must learn in order to fit it for the struggle for life, and a cat teaches her kittens how to do it. It may be that sometimes the whole future of a litter of kittens will be prejudiced through their having been taught a bad style of catching mice by their mother. So even among animals survival may depend on having been taught the right doctrines.

The processes of teaching and learning are obviously of immensely greater importance to humanity than to any animal, and the first thing to note is that in the matter of heredity education must rank as an *acquired character*, so that it does not come under the Mendelian laws. There can be no genes representing what one has been taught. Nevertheless, as I pointed out in the last chapter, the matter is rather more subtle than this bald statement might be taken to imply. To bring the point out, I will take an example which is deliberately exaggerated.

Suppose that there is an earnest believer who holds that acquired characters are always inherited perfectly. He ought to expect that the children of literate parents will be able to read without being taught, or that the son of a Latinist will spontaneously know Latin grammar; in fact he expects neither of these things, but only for the reason that every single child that has ever been born is an example to the contrary. Though he has to accept this disappointment, he will nevertheless make the best of things by claiming that the children of literate parents usually learn to read earlier and more easily than others do. In this he will often be quite right, but part of the reason is only loosely connected with heredity. In part the result arises from two very general characteristics of mankind, the tendency of the child to imitate what it sees going on round it, and the tendency of parents to want to teach their children. It is true that these are innate characters, so that they will fall under the biological laws of heredity, but they are too general to be invoked for the narrower purpose of the present argument. However, there is a more special application of the law of heredity which can legitimately be used. It is true that there is no gene conferring a knowledge of Latin grammar, but there may well be a gene conferring the type of brain which makes the study of things like Latin grammar congenial, and it is likely that parents, who have found Latin to their taste, will possess this gene, so that they are likely to pass it on to their children.

The question of the inheritance of intellectual qualities is matched by the inheritance of moral qualities, which have to be taught to the growing child in much the same manner, by precept and example. There still lingers in many quarters the conception of the perfectibility of mankind, and it is well worth considering what the biological theory of heredity has to say about it. In discussing the non-inheritance of acquired characters in the last chapter I cited the example of the training of dogs to do tricks, and I pointed out that the wise trainer would always use the dogs that learnt the tricks easily, and would get rid of those that were slow learners. I chose this example deliberately as being free from any ethical questions, such as must inevitably enter into all matters that concern humanity; but still it is interesting to see how it would apply to humanity, when considerations induced from ethics are for the moment forgotten. A philanthropic dictator wants to perfect the innate moral qualities of the human race; how should he go about it? Following the example of the dog trainer, he will devote all his attention to the good children, and he will neglect the worse ones, doing all he can to see that they do not succeed in life, and above all that they are not permitted to hand on their inferior qualities to later generations. Actually all too often philanthropic effort goes in exactly the opposite direction, into curing the faults of the worst, without recognizing that the acquired characters so induced are quite impermanent. In saying this I am thinking of the long-range policy, and I do not in the least want to belittle the self-sacrificing work that is done by so many noble workers in improving the conduct of the worse elements of the population. It may be justified as being a good in itself, and moreover the existence of criminals perturbs very seriously the life of the rest of the community, so that everyone benefits if this nuisance is removed. Still it is proper to note, that the policy of paying most attention to the inferior types is the most inefficient way possible of achieving the perfectibility of the human race.

Turning now away from these narrower questions of biological heredity, consider the larger question of how education, in the widest sense, has affected and will affect history. Every man builds up a world of thought, directing his conduct, which is partly formed from his own experience, but even more of it is acquired from his teachers, and in later life from friends and acquaintances, or from books. I shall use the word *creed* to denote a set of tenets acquired in this general manner. I mean the word in an entirely colourless sense, with no question arising of whether the creed is true or untrue, moral or immoral. It is merely a body of philosophical thought— whether it is reasonable or unreasonable philosophy— which is strongly held and used as a main guide to conduct. There are of course creeds held by single individuals, but naturally the important ones are those held by large communities. Such creeds have produced, and will again produce, enormous effects on human history, and their influence must be considered.

The first thing to notice about creeds is that they are inherited, but inherited according to curious laws quite different from the usual biological ones. A man is rather likely to hold the same creeds as his parents and relations, but no more than he is those of his teachers and his friends, whereas he has received his instincts and his inherent qualities from his ancestors and will share them with his blood relations; it will be pure accident if they happen to coincide with those of his friends. In this respect a parallel can be drawn between creeds and languages, for a man is likely to speak the language of his parents, but he is quite as likely to speak that of his unrelated friends. A language is a simpler thing to analyse than a creed, and so it may be useful to follow the analogy further. Up to a point the classification of languages resembles the classification of animals; both can be divided into varieties, species, genera and so on, and both gradually change their forms with the lapse of time, or they may branch out into several varieties, or they may become extinct. To this degree they are similar, but the resemblance fails if it is taken further. Thus the vocabulary of a language is sometimes a mixture derived from quite unrelated sources; among animals this would be as though a hybrid could be produced between a mammal and an insect. Again sometimes wholly new words arise, derived from no traceable source and presumably originating from the caprice of some inventor; for languages the principle *omne vivum ex vivo* does not hold. Much the same would seem to be true of creeds. Like animals, they could be classified into varieties, species and so on, and like them they often show progressive modification, branching and extinction; but on the other hand there have often been hybridizations between quite unconnected creeds, and sometimes wholly new doctrines have arisen with no evident parentage at all. In fact, a new kind of heredity has come into existence, of quite a different type from that affecting animals, but of great importance to mankind. It appears to me that what I may call the Natural History of Creeds would be a very exciting study and one meriting a great deal of attention.

I have neither the psychological nor the historical knowledge to study the natural history of creeds in detail, and I shall therefore be content with giving a few examples of what I am trying to express. I select for my starting point an avowedly trivial example, the creeds about what food should be eaten. Why is it that we eat the flesh of certain animals and not of others? If anyone in Europe is offered a dish of dog's flesh he will refuse it with something like horror, rationalizing his refusal perhaps by the explanation that the dog is a dirty animal. A Moslem will also refuse it; for him it would be immoral to eat it because it is prohibited by his scriptures. On the other hand, in certain parts of China, the dog is much prized as an article of diet. From what we know of other Chinese cooking this shows that the taste of dog's flesh must be excellent; nevertheless there are very few of us who could be persuaded to try the dish. Contrast this with the reactions to the pig. For the Chinese it is the most prized meat of all, for the European nearly as much so, even though the pig is proverbially regarded as the typical dirty animal, whereas for the Moslem it would again be immoral to eat it. Creeds about food are patently trivial, but this example does show nevertheless what a tremendously strong influence a creed has on our conduct.

Creeds about more important things naturally have a

very much greater compulsion. Those we hold firmly appear to us to have the inevitability of the propositions of formal logic. Anyone who does not happen to share our creeds is at the least regarded as an illogical fool, but more frequently as a perversely wicked person. It is this that has led to most of the terrible series of persecutions that have blackened the records of history.

Creeds often arouse the most fanatical devotion. It is enthusiasm for his creed that has created the martyr, and, if we happen to share his creed, the martyr is regarded as one of the noblest of humanity. But the matter is not as simple as that, for this judgment has usually been prejudiced by the fact that we do sympathize with the martyr's creed, and it is necessary to look at the subject without this prejudice. The martyr is driven to make the ultimate sacrifice by his enthusiasm for his creed, but this enthusiasm has usually been evoked by the counter-enthusiasm of his persecutors, the majority in power, who hold an opposite creed with equal fervour. For every man who is willing to die for his faith there will be ten men who are willing to kill for their faith. The ten feel that they are actuated by the same motive, the pure hatred of evil, as that of the martyr, and the main difference is only that for weak human nature the role of the persecutor is easier than the role of the persecuted. But that there is no very great difference between the two is shown by many examples in history, for when the persecuted party has gained the upper hand, it has usually indulged in counter-persecution on a scale equal to that which it had itself suffered. I have cited the history of persecution as an example showing how intensely important creeds are as influences on human conduct, and, in passing, another characteristic of them may be noted. This is that, though the infidel is hated, he is by no means so much hated as is the heretic. However, though such matters are interesting aspects of the natural history of creeds, for my present purpose it would be out of place to follow them further.

It will be noticed that I have not said anything at all about what is the fundamental question in regard to any creed, and that is whether it is true or false. For one who wants to believe in a creed its truth is all that matters, but it is not this that matters for my purpose. In the past there have been creeds, such, for example, as the belief in magic or divination, which have been very widely accepted, but we now know them to have been quite absurdly false. Yet they have exerted the very greatest influence on human history. The species *homo* has not changed, and there are still very many who are only too eager to believe in such things—not by any means all of them confined to the less advanced civilizations—and it must be expected that this tendency will continue to recur again and again. The question of the truth of a creed is therefore irrelevant to my purpose; what does matter is whether a creed, true or false, helps to the survival of its holder, and it is from this point of view that I shall try and study the natural history of creeds.

I can best illustrate the importance of creeds for survival by beginning with an example which is avowedly much over-simplified. One of the tenets of the Society of Friends is that it is wrong to fight. Quakers therefore would not be killed in war, whereas the believers of other faiths without this prohibition would lose a fraction of their numbers in every war. Religious faiths have a strong tendency to be adopted from the parents, and so in each succeeding generation there should be more Quakers in proportion to the rest of the population, and yet there is no difference whatever in the make-up of the genes of the body cells of the two types. This example illustrates the way in which a creed might affect survival; it has of course been much over simplified and it must not be pushed too far, for if it were carried to the extreme all the population would in the end be non-combatant, with no one to protect them from being destroyed by another race.

A much more important example is the ancestor-worship formerly prevalent in China. This imposed on a man the obligation to have a family in order that the worship of the shades of his ancestors might continue. With a population like the Chinese in which its poorest members are always living on the edge of starvation, there must have been a much greater chance of survival for the children of the abler people, so that the creed would have a strongly eugenic effect. Contrast this with the state of Europe in the Middle Ages. There it was the custom for many of the ablest people to go into the church, and so condemn themselves to sterility. Even if there was often laxity about the enforcement of celibacy, a priest's children would be illegitimate, and so would be handicapped instead of being favoured in their chances of survival. This difference of creeds goes with a remarkable difference of histories, and it may well have been an important contributing factor to the difference. Both the Chinese and the Roman empires were attacked at various times by barbarians, and whereas the Roman Empire was so disrupted that it took nearly a thousand years for civilization to return to it, the Chinese Empire absorbed its Mongol conquerors after only two generations. Is it not probable that it is largely on account of the creed of ancestor-worship that the Chinese civilization has been the one in the world that has shown the most continuity, and the one that now has a fifth of the whole human race?

In the study of the natural history of the creeds of the past, it is inevitable to consider the religious creeds in particular, both because they are the ones that have roused the passions of humanity far the most, and also because we have much fuller records of them than of any others. It must be remembered that from the present point of view the question of the truth or falsity of a religion is not directly relevant; the only question is its influence on the history of the human race. In each of the great religions of the world, however different their purely theological doctrines may be, there has been a general ethic which has exerted a steady and beneficent influence on its believers. The ethics of the different religions have not been very different, since their main aim is to inculcate the social virtues which are essential if life is to be tolerable in any community, large or small. For example, a virtuous Christian and a virtuous Moslem will have very similar standards of conduct dictated to them by their very different religions, and these standards will be hardly different from those dictated by the Confucian philosophy. The influence of these ethical principles has been immense, and, if I do not discuss them further, it is not for want of recognizing this influence; it is because they do seem to have worked out to the same consequences no matter what the religion from which they started, whereas an objective study of creeds must be primarily concerned, with the different consequences that they may have produced. A greater interest attaches to the aspects of religious creeds where enthusiasm or even fanaticism has entered, because it is these enthusiasms that have been responsible for the most striking events in history. For the rest of this chapter I shall use the word creed therefore in this more restricted sense.

In this narrower sense, creeds are almost like living things, possessing a course of life, from birth through maturity to death. The analogy is imperfect in one respect, since all too often, after the general enthusiasm for the creed is dead, there is left behind a minority, a sort of fossilized rump, which continues to hold the doctrines of the creed for centuries afterwards. With this qualification, and perhaps with other exceptions, it would seem a rough generalization that creeds tend to live for two or three centuries, or to express it in biological measure, for not more than about ten generations.

Consider some of the creeds that have flourished excessively inside the Christian religion. In the fourth and fifth centuries there were fanatical creeds associated with metaphysical questions of the nature of the deity, and men were ready to die, and to kill, for the sake of subtle questions, incomprehensible to us now, connected with the nature of the Trinity. As time went on this creed reached its old age and tended to become a political persecution, not of individuals, but of whole nations such as those that had adopted the Arian heresy. Then again in the eleventh century there were the Crusades, a more fitting creed for the semi-barbarous peoples of western Europe. They lasted about two centuries, and also degenerated in the end into a political instrument for rival Christian parties, who by that time had little left of the original enthusiasm against Islam. Then there is the Reformation, which started towards the beginning of the sixteenth century. Some may hold that we are still too near the Reformation to pronounce an opinion on its present vitality, but it is certain that its colour had very materially changed within little more than a century, for the Thirty Years War was a war for power, not for religion, even though it was largely between Protestants and Catholics. It would be most interesting to study whether there were similar growths and decays of creeds in Islam or Buddhism, and also to study the behaviour of such creeds as philosophical Stoicism, which never provoked the same fierce fanaticism as did the religious creeds.

Another feature of creeds seems to be rather general. Though the majority of a population, say something like nine-tenths, accept their creed implicitly and regard it as part of the law of nature, there is always a small minority who do not. Most people—call them the sheep —follow the ideas of their leaders unquestioningly, but this minority—the goats—goes by contraries, and disbelieves anything just because those around them believe it. The goats are often not very pleasant people, but they are usually above the average of intelligence. It is probably the corroding influence of the goats that gradually saps the vitality of a creed by its cumulative infection, and indeed there may well be a proportionality between the number of goats in a community and the life span of the creed of the sheep in that community.

In future history the constancy of human nature makes it certain that man will continue to be dominated by enthusiasm for creeds of one kind or another; he will persecute and be persecuted again and again for the sake of ideas, some of which to later ages will seem of no importance, and even unintelligible. But there is one much more valuable aspect of creeds that must be noticed. They serve to give a continuity to policy far greater than can usually be attained by intellectual conviction. There are many cases in history of enlightened statesmen who have devoted their lives to carrying through some measure for the general good. They may have succeeded, only to find that the next generation neglects all they have done, so that it becomes undone again in favour of some other quite different way of benefiting humanity. The intellectual adoption of a policy thus often hardly survives for more than a single generation, and this is too short a period for such a policy to overcome the tremendous effects of pure chance. But if the policy can arouse enough enthusiasm to be incorporated in a creed, then there is at least a prospect that it will continue for something like ten generations, and that is long enough to give a fair probability that it will prevail over the operations of pure chance. Thus a creed may have the rudiment of the quality, possessed by the genes of mankind, of being able to produce a permanent effect on humanity.

If the history of the future is not regarded as the automatic unfolding of a sequence of uncontrollable events —and few of us would accept this inevitability—then anyone who has decided what measures are desirable for the *permanent* betterment of his fellows will naturally have to consider what is the best method of carrying his policy through. There are three levels at which he might work. The first and weakest is by direct conscious political action; his policy is likely to die with him and so to be ineffective. The second is by the creation of a creed, since this has the prospect of lasting for quite a number of generations, so that there is some prospect of really changing the world a little with it. The third would be by directly changing man's nature, working through the laws of biological heredity, and if this could be done for long enough it would be really effective. But even if we knew all about man's genes, which we certainly do not, a policy of this kind would be almost impossible to enforce even for a short time, and, since it would take many generations to carry it through, it would almost certainly be dropped long before any perceptible effects were achieved. That is why creeds are so tremendously important for the future; a creed gives the best *practical* hope that a policy will endure well beyond the life of its author, and so it gives the best practical hope that man can have for really controlling his future fate.

VII

MAN—A WILD ANIMAL

IN the past two chapters I have examined different aspects of the nature of man. In the first he was regarded just like any other species of wild animal, while in the second some of his social qualities were considered, which might not be regarded as those of a wild animal. Civilization might, loosely speaking, be counted as a sort of domestication, in that it imposes on man conditions not at all typical of wild life. It might then at least be argued that it is a false analogy to compare man to a wild animal, but that he should rather be compared to one which has been domesticated. I shall maintain that this analogy would be false, and that man is and will always continue to be essentially a wild and not a tame animal.

Before coming to this main theme it is important to notice that, if it were admissible to regard man as a domesticated animal, the whole time-scale of history would have to be radically altered. Thus though the geological evidence shows that it takes a million years to make a new wild species, we know that the various domesticated animals have been created in a very much shorter time. For example, the ancestors of the greyhound and the bulldog of ten thousand years ago would probably have been quite indistinguishable. If then man's characteristics could be similarly remoulded in so short a time, the whole future of history might be radically different. It would become impossible to forecast man's future after as short a period as ten thousand years, hardly longer than the span of known past history, instead of the million years which holds if he is a wild animal.

In the first place, it is necessary to be clear as to what is meant by a wild or a tame animal. We are apt sometimes to call an animal wild because it is dangerous to man, and to call it tame because it is harmless, but this is a slovenly way of speaking, and here I shall use the word "tame" simply as a synonym for "domesticated" which I think is its true meaning. A tame animal then is one that does the will of a master, and the savage watch-dog, trained to bite all intruders, is tamer than the friendly terrier which sometimes slips away to do its own private hunting. All tame animals owe their qualities to centuries of selective breeding, and it must always be remembered that the changes made in them owe nothing to the inheritance of acquired characters, but are due to the selection for breeding of those individual animals which show to the highest degree natural characteristics useful to their masters.

A chief feature in domesticated animals has been the creation of a great variety of breeds, each specialized for some particular purpose, either practical or aesthetic. Each breed far excels its wild ancestry in the quality for which it has been bred, so that race-horses run faster than wild horses, dairy cows give much more milk than any wild cattle, and the sheep-dog has even been bred to do skilfully the exact opposite of what the ancestral wolf would have done. Now human families often show special qualities in which they excel their fellows, and in some cases these qualities seem to be hereditary—witness the musicians of the Bach family. If man is really a tame animal, there is no reason why breeds of man should not be created, say breeds of mathematicians or of professional runners, who should possess gifts far beyond anything we now know, and far beyond anything that their fellows could compete against. Certainly at the present time mankind is very far from this, but that would not exclude the possibility in the not so very distant future, if man really were a tame animal. I shall consider this question of special breeds later in the chapter in more detail; all the evidence seems to show that they will not arise, but to see this clearly, it is best to return to the prime feature of tameness, obedience to a master.

It is obvious that we in this country, with our passion for freedom, value wildness very highly, whereas in some lands, where the population are content to live under a much more strictly controlled rule of discipline, tameness may be more nearly acceptable. This question of taste is irrelevant however, for it might be that a tame race could achieve so much higher a degree of efficiency that it could master the wild ones, and so reduce them also to a state of tameness. I am going to maintain that this cannot happen, in that man is untameable. The reason involves a feature not often present in scientific arguments, and I will venture to introduce it by means of a fable.

There was a man who was endowed with very great intelligence, very great wealth, deep scientific knowledge and a benevolent wish to improve the lot of mankind. He also knew himself to have the gift of longevity, so that he had the prospect not only of starting on his beneficent plans, but of seeing them through to their final accomplishment.

Now it chanced that about this time there came up for sale a large uninhabited island, enjoying a climate in every way fitting it for human habitation. He purchased the island, and persuaded a large number of his friends and admirers to come with him, and to settle there and live under his direction.

The director first made a thorough study of his people's natural gifts and capacities, and then he set each man and woman to the work for which they were specially fitted. Artisans were chosen who had both skill in craftsmanship and a liking for their trade, domestic servants who had a passion for cleanliness, and cooks who were really interested in the taste of the food they produced.

He chose for schoolmasters those who could best inspire their pupils; his professors and research workers were selected because they showed the highest flights of scientific imagination; his lawyers possessed the greatest subtlety in argument; his civil servants and industrialists were those who were gifted with the highest qualities of administrative ability.

Nor did he neglect the other sides of his people's interests, for he selected those who were gifted at painting and music and poetry and encouraged them to practise their arts. He also had actors and actresses of great charm and beauty, and athletes who could run very fast, jump very high, or guide a ball with remarkable accuracy.

Having laid the foundation of his plan in this manner, he persuaded them all through his dominating personality—nor, be it said, was compulsion entirely absent—to mate together in such a manner that the various special gifts of each group were conserved and enhanced.

Those of the settlers who proved unsatisfactory in mind or body were gradually eliminated, not by exile or punishment, but merely by forbidding them to enter into fertile unions.

At the end of ten thousand years he had achieved results in humanity even more remarkable than those that have been already achieved in this span of time in our domestic animals. His actresses were of surpassing beauty, and his athletes, whose limbs had attained very highly specialized proportions, were so persistently victorious that international contests became impossible. All the really ingenious machines in the world were contrived by his engineers, and in their construction his artisans were pre-eminent. His diplomatists could always get their way with the diplomatists of other countries. His research workers made remarkable progress in the development of scientific knowledge—though it was perhaps not often they who started any wholly new subject.

The director had produced this surpassing improvement in the quality of his subjects in the course of ten thousand years, when it was revealed to him that his life was nearing its end. And now he saw that his work had been in vain, because he had made no provision for a successor to himself.

He had moulded his subjects so that they fulfilled their tasks superlatively, because he could look at those tasks objectively, but his own task he could only know subjectively, and the prescriptions he had used for the others were without avail.

He had tamed men into being domestic animals, but he could not tame anyone into being a director, because a director must be a wild and not a tame animal.

Though this has only been presented as a fable, the experience with domesticated animals does show that the most astonishing improvements could be made in the various human faculties, if a similar course of continuous selection could be applied to man over as long a period of time. The trouble is that for man this is not possible, because he has got to apply the selection to himself, and that means that it is not merely a different problem, but a wholly different kind of problem. There is a fundamental difference between the subjective and the objective. Scientific progress has always succeeded only by regarding its themes of study objectively; even in the field of psychology progress has mainly come by the study of the minds of others, that is to say objectively, instead of by following the old barren course of introspection. The most severe critic of his own conduct can never judge his actions as if they were someone else's, and the selective breeding of other types of people would be no guide at all in the breeding of his own kind.

If the director had foreseen his death, he would have

tried to produce a successor to himself. Since his profound belief in heredity had been so fully confirmed by
the remarkable changes he had made in his subjects, he
would naturally expect that it would be one of his own
sons that would be best fitted to succeed him, but his
difficulty would be just the same if he were trying to
find a successor elsewhere. The matter is on quite a
different footing from all his other decisions. For the
others he could say: "I have improved all our breeds, by
seeing which son improved on the qualities of his father.
That is why I select you." For his own successor the utmost he could say would be "I am selecting you in the
hope that you may be a better director than I have been.
But I have no idea how you will set about it, since, if I
had known what I was failing in, I should have set it
right myself." The targets in the two statements are
quite different, for in one he knows what he is aiming at,
in the other he does not. In one case the target is to make
the man better, in the other to hope to make him as
good. One is the systematic breeding of tame animals,
the other the unsystematic method of nature in the
breeding of wild animals.

This point is so important that before following it to
its conclusion I will give another example, which has
the advantage of not being fabulous. In their studies of
how to improve the human race the eugenists have very
naturally considered both ends of their problem, the
increase in the good qualities of humanity and the
elimination of the bad qualities. Their chief effort has
gone, quite rightly at first, into the easy part of the
problem, and they have spent most of their energy in

pointing out the disastrous tendencies of the present policy of directly encouraging the breeding of the feeble-minded. This is undoubtedly useful work, but it is comparatively easy, since these feeble-minded can be regarded objectively by their superiors, and so might become amenable to the same sort of control as is applicable to domestic animals. This restraint of the breeding of the feeble-minded is important, and it must never be neglected, but it cannot be regarded as a really effective way of improving the human race. If by analogy one wished to improve the breed of racehorses, one might accomplish a little by always slaughtering the horse that finished last in every race, but it would be a much slower process than the actual one of sending the winner to the stud farm.

Conscious of this criticism, eugenists have often attempted to define what are the good characteristics which should be positively encouraged, instead of only the negative ones that must be discouraged, but the results are disappointing. Lists of meritorious qualities such as good health, good physique, high intelligence, good family history, are compiled, and those possessing them are told that they should breed, but the statements lead nowhere in practice, for no one can be expected to assess his own merits and demerits in a balanced way. How, for example, is a man to weigh his own good health or good ability against a heredity made dubious, say, by an uncle who was insane, or again how is he to strike a balance between considerable artistic gifts—as he thinks—together with a good family record, but quite bad health. It is clearly beyond anyone to decide these things for himself, and even then the matter is only half settled, since similar judgments are needed for both partners to the marriage. However helpful the literature may be which can be consulted, it is evident that subjective judgments on such matters are too difficult; with the best will in the world they would very often be made wrongly, because, however sincerely he tries, no man can be a good judge in his own case.

The only imaginable way of overcoming these difficulties would be to set up a class of consultants who would prescribe what marriages were eugenically admissible and how large the consequent families should be. But this does not solve the difficulty; it only pushes it back a stage, for it leaves unanswered the question who are to be the consultants, and what principles are to guide them in settling the values of the different qualities of mankind. It comes back to just the difficulty I described in my fable, that a tame animal must have a master, and that therefore though it might conceivably be possible to tame the majority of mankind, this could only be done by leaving untamed a minority of the population. Moreover, this minority would have to be the group possessing the most superior qualities of all.

These examples suggest the impossibility of taming mankind as a whole, but before accepting the principle fully, it is proper to examine a case where the exact contrary has happened; this is in the insect civilizations of the ants or termites. In applying the same term, civilization, to both ants and men, it is hardly necessary to say that I am drawing an analogy between things which are really of a very different quality. All species of ants live in cities, and some species have developed agriculture, others animal husbandry; but all these practices are purely instinctive and individual to each species. On the other hand human civilization is an acquired character, based on education, and so is not inherent in man's nature. Nevertheless it may be worth while to follow out the analogy a little further. Admitting the different sense of the words, it may be said that all species of ants have made the third revolution, the invention of cities, that some have made the second, agriculture, none the first or fourth, fire and science; but they have all added another revolution of their own, the complete control of the problem of sex. The ants' nest has no rulers at all, for the queen is hardly more than an egg-laying mechanism, and they seem to get on perfectly well without civil servants or lawyers or captains of industry.

Why cannot man set up a community like an ants' nest? This would be the ideal of the anarchist, and hitherto it has held no promise at all of success, but with the help of recent and probable future biological discoveries, some sort of imitation by man of the ants' nest cannot be quite excluded from consideration. Thus the control of the numbers of the two sexes may become possible, and with the knowledge of the various sexual hormones it might also become possible to free the majority of mankind from the urgency of sexual impulse, so that they could live contented celibate lives, instead of the unsatisfied celibate lives that are the compulsory lot of such a large fraction of the present population of the world. If these discoveries should be made—and this is really by no means impossible—man would be able to carry out the sex revolution which is the typical characteristic of the insect civilizations. The detail would of course have to be quite different, for instead of one queen there would have to be large numbers of fertile women to renew the population, whereas there might be one king, literally the father of his country. Also it is probable that on account of their greater physical strength, it would be the men who would be the workers.

Such an organization is certainly repellently unattractive to most of us—perhaps excepting some of the autocrats of the present world—but it is not this that excludes the possibility of it. There is no danger whatever of its happening, because of the inherent difference between vertebrate and insect, for the vertebrate is so very much more flexible than the insect in its behaviour. Most insects simply die if placed on an unfamiliar food plant, whereas the vertebrate will always try experiments if its normal diet fails. An insect can be used to prey on and destroy another one that has become a pest, and, when it has done so, the predator will die of starvation; in the same role a vertebrate predator would not die, but would start to destroy some other, perhaps beneficent, species. Now of all vertebrates man is preeminent in his willingness to try experiments, so that it is inconceivable that he should settle down into the inflexible unquestioning course of life that is typical of an insect. It would call for a quite radical change in his whole nature. It would not be a mere change into a new species of *homo* that would be needed, nor even a change into a new genus or family or order of the mammals. It would have to be a fundamental change into a new phylum of the animal kingdom, and that would not take a mere million years, but many hundreds of millions of years.

There is no prospect of man's nature imitating an insect's, but it is much more nearly imaginable that his development should go, like that of the dogs, into a set of breeds each specialized for a particular purpose. We all of us know of whole human families which possess gifts specialized in some particular direction, and if the specialization were narrowed and the gifts improved till all competitors were surpassed, such a family would have turned itself into a breed. But all past history contradicts this tendency, for it suggests that wherever there have been such groups they have not increased further in their specialized skills, but that after a very few generations they have tended to merge back into the general population. I will give some examples, though my knowledge of history is hardly deep enough to cite them with any confidence.

A first example may be drawn from the sanctity of royal blood, which has been a prevalent idea in many countries, and which would give opportunity for the in-breeding that is essential for the production of a specialized breed. The most extreme case is that of the dynasty of the Ptolemies in Egypt whose blood was counted as so sacred that the reigning house had to be perpetuated by brother-and-sister marriages. Biologists no longer now regard close in-breeding as necessarily deleterious, but still the possibility of its evil effects might throw doubt on any positive conclusions we could draw from the Ptolemies. But the only conclusions that can be drawn are entirely negative; the record of the dynasty is not very impressive, it is neither much better nor much worse than that of other dynasties that had not been in-bred, and in the end it collapsed, as did the other dynasties, under the irresistible might of the Romans. Neither in this extreme case, nor in other more modern ones, is there any sign of a tendency for a breed to arise that is specialized for kingship.

It might be contended that the number of individuals in reigning houses is too small to give rise to a breed, and my next example concerns a much larger population. It is the military caste of German nobles in the sixteenth, seventeenth and eighteenth centuries. Whatever the extra-nuptial habits of this caste, its marriages were most strictly regulated, so that it might have provided the starting point for a specialized breed. It is undoubted that these noble families provided some very good generals; this was inevitable since in their own country they had the monopoly of the officer ranks, but they were not conspicuously better than other generals who did not belong to the caste. In making the comparison it might be argued that Louis XIV's generals should be excluded, as themselves belonging to a similar caste, but the German military nobles were also far excelled by others, such as Marlborough, who, though of gentle birth, certainly was not drawn from the military caste. Furthermore, if this caste had shown promise of turning into a breed, it should have produced better generals at the end

of three centuries than at the beginning, whereas if anything they had degenerated. Excluding Napoleon from
comparison, as an exception to all rules, they showed no
marked superiority over his marshals, who came from
all classes of society. In three hundred years this caste
certainly showed no signs of turning into a specialized
breed.

A more striking example is the caste of the Brahmans
in India, because its purity has been preserved over
many centuries by the religious sanctions of their creed.
They have the advantage of being much more numerous
than the castes I have cited hitherto, and they have
very certainly played a conspicuous part in the history
of India, but they show none of the tendency to an
increase in specialization that should characterize the
creation of a breed. Since they were never a military
caste it is not surprising that many of the reigning houses
of India are not Brahmans, and the priestly function of
the Brahman would more naturally destine him to play
the part of philosopher or intellectual. Now it is true
that in the modern Indian universities a considerable
fraction, perhaps a majority, of the distinguished professors belong to this caste, but still there are quite a
number of others as distinguished who do not. It seems
at least doubtful whether in this there is any real
difference between India and Europe, for in Europe also
a very considerable fraction of the intellectual life is
contributed by what might be called the hereditary
middle classes, that is by families which have continued
through succeeding generations to show a general intellectual ability, though they are in no sense an exclusive

caste. Once again, in these exceptionally favourable circumstances, there is no sign of the Brahmans turning from a caste into a breed.

This is the place to refer to the case of the Jews, because though very superficially it might be thought similar, it is really quite different. It is true that for centuries they practised the close in-breeding that would be needed for the creation of a specialized breed, but the point is that they have shown no signs of becoming specialized, for there have been Jews who have excelled in every one of the arts and sciences of civilized life. One of their distinguishing features has perhaps been that they were earlier adapted, than the more recently civilized western Europeans, to the crowded life of cities, but this is not so much a specialization as an adaptation in which they have anticipated the others. In the course of the centuries their race has had one great advantage, though they would certainly very willingly have foregone it. This is their long history of almost continuous persecution, and it is tempting to believe that this has been an important factor in giving them their high qualities. In order merely to keep alive, a Jew had to show intelligence more frequently than did the surrounding peoples, and this intelligence was gradually incorporated in his heredity. But in all this there is no sign of specialization; at most it is a more complete adaptation to the crowded life of cities than has been hitherto shown by the rest.

All these examples confirm that there are specialized abilities in some of humanity and that they are often hereditary, but they hold out no expectation that the specializations will spontaneously become narrower or that they will rise to higher levels, which is what they would have to do if man were destined automatically to branch out into breeds as distinct as those of the domestic animals. There may be those who will regret that man will not attain these pinnacles of specialization, but the failure is inevitable. In order to create such specialist breeds there would have to be a master breed at the summit, and this would be a totally different kind of thing from all the other breeds, because it would have to create itself.

At every turn the argument leads back to this question of the master breed. Nothing can be done in the way of changing man from a wild into a tame animal without first creating such a breed, but most people are entirely inconsistent in their ideas of what they want created. On the one hand they feel that all the world's problems would be solved if only there were a wise and good man who would tell everybody what to do, but on the other hand they bitterly resent being themselves told what to do. As to which of these motives would prevail, it seems at least probable that it would be the resentment, so that if the breed should arise in any manner, it would be extirpated before it could ever become well established. It is, however, imaginable, that there might be a part of the world in which the breed was accepted, and that this part should gain a superiority over the rest of the world, because it could develop various suitable breeds of specialists under the control and direction of the master breed, and by the exercise of the skills of these specialists it might overcome the other nations. So it is appropriate to look a little further into the matter.

Imagine that through new discoveries in biology, say by suitably controlled doses of X-rays, it becomes possible to modify the genes in any desired direction, so that heritable changes can be produced in the qualities of some members of the human race. I may say I do not believe this is ever likely to be practicable, but that does not matter as far as concerns the present argument. The first success might be in some physical attribute, for example, by making a breed with longer and stronger legs so that it could jump a good deal higher than anyone can at present. But passing to more important matters, there might be created a breed which could think more abstractly, say a breed of mathematicians, or one that could think more judiciously, say a breed of higher civil servants. These would be of great value, but they would not be the master breed, and the question arises of a more precise prescription for what the qualities of the master breed are to be.

It is usually best to build on what one already has, rather than to start from nothing. So the natural procedure would be to begin with existing rulers, since these have already established themselves as acceptable to at least a good many of their fellow creatures. One would collect together, say, a hundred of the most important present rulers—among them of course should be included a good many who exert secret influence without holding any overt office—and tell them to get on with the business of settling what the master breed should be. It is impossible to believe that any such body of men would ever reach agreement on any subject whatever; so this plan fails.

In the search for the qualities of the master breed the next idea might be to appeal to the wisdom of our forefathers. Plato in his *Republic* devotes much attention to this very subject. Why not then find a Plato, give him his group of recruits, and let him educate them for thirty years according to his prescription—though perhaps fortifying it by the findings of modern educational theory; the result should be the master breed. But this will not do either, for Plato was not educating the master breed, he was educating the civil servant breed. It is not about these that there is any difficulty; it is the finding of someone to fill the role of Plato himself. It all comes back to the point that we do not know in the remotest degree what we want; for I do not count as an answer the one that would usually be proposed, which would be that the type required should be good and wise, while at the same time showing a special favour for the particular enthusiasms of the proposer. The reason for the impossibility of making a prescription for the master breed is that it is not a breed at all; to call it so is to change the sense of the word. Breeds are specialized for particular purposes, but the essence of masters is that they must not be specialized. They have to be able to deal with totally unforeseen conditions, and this is a quality of wild, not of tame, life. No prescription for the master breed is possible.

In these considerations I have been assuming the licence of supposing that we might be able really to change human nature in a heritable manner, and this is far beyond all probability. Returning now to more practical considerations, there seems no likelihood whatever of a master breed arising. All through history the most formidable difficulty of every ruler has been the selection of his successor, and the best intentions have been nearly always disappointed. Indeed it is notably surprising how very seldom the choice has been well made. The immediate cause of these failures has been the difficulty of the subjective judgments on the basis of which the choice must be made, but fundamentally they have arisen from a cause in the deep nature of mankind. Of all animals man is the most ready to try experiments and there are always candidates—far too many candidates—who regard themselves as fit members for the master breed. This quality is a characteristic of a wild animal, and it will always prevent man from domesticating himself. He will always prevent the creation of the master breed, through which alone the rest of man could be domesticated. The evolution of the human race will not be accomplished in the ten thousand years of tame animals, but in the million years of wild animals, because man is and will always continue to be a wild animal.

VIII

LIMITATION OF POPULATION

IN the past chapters I have discussed some of the basic qualities of man, which show that in general he behaves, and will continue to behave, like a wild animal. Consequently he will obey the law of nature by multiplying up to the limit of subsistence, and there will therefore normally be a marginal group of humanity living—and dying—at the starvation level. This is the old threat of Malthus, and there are many people at the present time who are very conscious of it. But whereas Malthus could only express the hope that man would learn voluntarily to restrain himself from multiplying up to the limit, it has been found since then that in some countries, and those among the most prosperous, the populations are spontaneously becoming stationary or even decreasing. If then, it is argued, some countries have created conditions which automatically solve the problem of over-population, why should not these same conditions produce the same results in all countries, and then we could all be comfortable together. I am going to maintain that, though such limitations of population may recur locally from time to time, the condition is essentially an unstable one and contains the seeds of its own destruction. Any country which limits its population becomes thereby less numerous than one which refuses to do so, and so the first will be sooner or later crowded out of existence by the second. And again, the stationary population is avoiding the full blast of natural competition, and, following a universal biological law, it will gradually degenerate. It is impossible to believe that a degenerating small population can survive in the long run in a strongly competitive world, or that it can have the force to compel the rest of the world to degenerate with it.

The present spontaneous decline in fertility affects many of the most prosperous countries, and it is a phenomenon that was hardly foreseen even fifty years ago. Man has not any very strong procreative instincts; in the uncivilized state that did not matter, since his very strong sexual instincts sufficed to maintain the race, but our present economy is so organized that there are great handicaps against large families, and many compensations for those who are either intentionally or unintentionally sterile. In past periods of high prosperity, there was often a similar state of affairs, but the development of easy methods of birth-control is a new factor of great importance, which seems to have upset the balance by making it so very easy to be childless, or to have such a small family that the population is not maintained. I believe it is disputed how important the various contributory causes may be, but it is indisputable that a considerable fraction of the population find it both easy and convenient to contribute less than their share to the next generation, and this fraction is specially the one enjoying the highest prosperity.

It is convenient to have a short phrase to describe this

state of affairs in which prosperity produces childlessness, and I shall characterize it by saying that the prospect
of owning a motor-car is a sufficient bribe to sterilize
most people. I do not apologize for calling it sterility,
for though the term is often used to imply a physical
incapacity that is held in contempt, it is, biologically
speaking, immaterial whether the incapacity is forced or
voluntary. In my phrase the motor-car is of course only
metaphorical, as a symbol of the sort of level of prosperity that tends to be associated with small families or
childlessness; and it is being found that as prosperity
spreads downwards in the social scale, so the families
tend to become smaller there too. It would be difficult
to say which is cause and which effect, for children are
an economic disadvantage, so that their presence lowers
their parents' prospects, and on the other hand the ease
and comfort of existing prosperity discourages the
creation of children. To see the consequences of this
state of affairs, I shall look more closely at the way things
have been going in this country during the past century.

In studying the trend of our population, if the study
is to be any use at all, it is necessary to adopt some
standard of values for the different constituents of the
community. There is at present current in some quarters
an equalitarian trend of opinion which is quite danger-
ously unsound; it is the type that condemns all eugenic
views on principle, presumably because they conflict
with a dead level of equality. It tries to prejudice the
case in advance by stating that the eugenist rates the rich
higher than the poor, without any examination what-

ever of the very different things that he really does claim. It may therefore be well for me to elaborate the point.

In discussions on social policy, in which these criticisms are expressed, lip-service is paid to the doctrine that all men are of equal merit, but it is to be noticed that such statements are usually reserved for general argument, and that those who make them pay the most jealous attention to the comparative merits of individuals, when it is a question of making an appointment to any important post. It is the man who is believed to be the ablest who is promoted, and this brings to him some increase in reward, which may be a higher salary, or perhaps some other mark of civic recognition, which will be valued by the recipient, and which would also have been desired by his unsuccessful rivals. There is certainly a great deal of injustice in the world, in that there are many people whose real ability is never discovered, but it is hard to believe that promotions are more often made wrongly than rightly. So it is surely a justifiable claim that those selected for promotion are rather more likely to have superior qualities than those who were not so selected. Now man, like every other animal, does tend to pass on his natural qualities to his offspring; there is no certainty about it, but there is a somewhat better chance that the sons of the promoted candidate will be abler than those of his unsuccessful rivals. Since there will always be need for as many able people as possible, the encouragement of the promoted man to have children increases the chance that we shall find them in the next generation. The argument may be pushed further still. There is a good deal of evidence that some men's ability is more intimately incorporated in their heredity than it is for others. Thus there have been men of pre-eminent ability, risen from the ranks, whose descendants have sunk back in a generation or two, whereas there are families where generation after generation goes on producing men of very good ability. Clearly the probability of producing able men is rather greater in a family that has shown that it can do so over several generations.

The argument that the eugenist rates the well-to-do highly is quite true if it is read in these terms, for the well-to-do are rather more likely than others to possess the quality of hereditary ability through having shown it in several generations. Note also that an opponent of this view does not really upset the argument by maintaining that the wrong people are always promoted. He wants other types to be promoted presumably because he admires their qualities more, but then he will surely want those other qualities to be perpetuated through heredity; it is not the eugenic side of the argument that he wants to upset, but the social side. All these matters, both of achieving success and of heredity, depend of course very much on chance, for very often the sons of the successful are inferior to those of an unsuccessful rival, but there is no justification for neglecting chances because they are not certainties. After all the whole world depends very much on chance, and it is the part of the wise speculator to recognize which chances are likely to make the odds most in his favour, and he will take these chances even if they are only slightly more favourable to him.

If then I may appear to be regarding the actually successful members of society as more valuable than the less successful, it is not because I do not recognize that there are a great many stupid rich people, or that there are many of superior merit who have been missed. It is because I believe that I shall tend to find rather more ability among those, some of whose ancestors have proved they possessed it, and in dealing with probabilities I want to have the chances in my favour as much as I can, even if the gain in the odds is not very great. In what follows, then, I shall be implying this train of thought; it would be too tedious to have to repeat it all the time. There is one further point to be made. The judgment by success is one between men competing against one another, and so it can only be applied when they belong to the same community. It provides no guide whatever to the respective merits of separate peoples, whether the comparison is made of the whole peoples, or of individuals drawn from each at some corresponding level.

Consider now the history of Britain since the Industrial Revolution. The large increase of population started in about 1800, and this signifies that the rigour of natural selection began to be eased at that time. There was still of course a great deal of infant mortality among all classes, but it was probably a good deal less among the well-to-do, though even theirs would be regarded as quite shocking by modern standards. Many of the poor led lives of oppressed squalor, and no doubt often lived not very far from the starvation level, but they did not actually starve and anyone who could survive childhood had a nearly equal chance of himself having children, no matter from what rank he came. Any difference there was would still favour the prosperous, though to a less degree than before the industrial revolution.

The situation had changed radically before the end of the nineteenth century, on account of a variety of causes, of which the comparative importance is still in dispute. One was the greater insistence of public opinion on sexual morality, and another, probably the most important of all, was the growth of the practice of birth-control. It might more doubtfully also be argued that the spread of comfort and the rise of living standards has provided pleasures to rival those connected with sex, but this is not very convincing, since it is certain that in other countries and at other periods of history the growth of luxury has had the opposite effect. But whatever the causes, it is indisputable that the more prosperous members of the community are not producing their share of the next generation, so that selection is now operating against the prosperous. As an example, if the list of candidates is examined, who are applying for any office of high or even mediocre importance, it will be found that something like nine-tenths of them have either no children, or one, or two. Of course, if everyone had exactly two children, and both these children married and had exactly two more, the population would be exactly steady, but as things are, it is a fair guess that, in each thirty years of a generation, this part of our population is reducing itself to something between a half and two-thirds. This signifies that within a century, there will at most be quarter as many people of this type as

there are now. There will of course be some compensa-
tion by the rise from other levels, but, as I have pointed
out, to found our hopes on them is to take a worse instead of a better chance. The whole thing is a catastrophe
which it is now almost too late to prevent. If what I have
called the bribe of the motor-car is what is needed to
persuade the world to limit its population, then it is
certain that the first countries to accept the bribe are
committing suicide.

This catastrophe must be a principal factor in the immediate future of our country, and as such it concerns
us more than anything in the distant future, but in this
essay I am concerned with the distant future, and not
with the immediate troubles of our country, so here it
only plays the part of an example of what happens when
a country succeeds in avoiding the Malthusian threat of
over-population.

The tendency of civilized life to sterilize its ablest
citizens is by no means confined to this country, but is
the experience of nearly all countries which enjoy even a
passable degree of prosperity. It is perhaps more marked
now than ever before, but it has certainly occurred at
other periods of history. For example, the earlier Roman
emperors were continually in difficulty because of the
extinction of the senatorial families, which were the
class whose administrative ability had been so largely
responsible for the creation of the Roman Empire. It
would seem that, then as now, it was just those whose
type was most needed who were the first to limit their
families. Then as now, the prosperity induced by civili-

zation gave not merely a security of life that annulled the effects of natural selection, but it actually went in the opposite direction, in that the less valuable parts of the community became the most efficient in survival.

Another example of the consequences of family limitation may be cited, but I do so only very tentatively because I have not been able to gather full information about it. Of the Polynesians many lived in small, nearly isolated, communities on islands, and these succeeded in developing a manner of life which seems to have avoided the harder features of over-population. It was done by the sanction of various rituals, and by social habits not all of which we should commend, but that is no matter. On their first being discovered they could be held up to the world as having solved the problem of how to live an idyllic life of Arcadian simplicity. But since then it has been found that Arcadia cannot endure in a cold world. The Polynesians are not in the least inferior to other races in ability or intelligence, but they do not seem capable of competing against them in survival. Thus in Hawaii, after the short span of a century, they are already very much in a minority compared with the newly immigrant Chinese and Japanese. Contrast this with the colonization of Africa, where the effect has usually been a rapid increase in the local population. Furthermore, the Polynesians themselves furnish one striking example to the contrary. When the Maoris came to New Zealand, they could expand into an almost unlimited area, and there was no need to limit population; and the Maoris have most certainly not gone the way of the other Polynesians. Through insufficient knowledge I can only cite this example of the Polynesians very tentatively, but it does seem to show that a race adapted to limiting its population cannot compete against others which have not been similarly adapted.

Those who are most anxious about the Malthusian threat argue that the decrease of population through prosperity is the solution of the population problem. They are unconscious of the degeneration of the race implied by this condition, or perhaps they are willing to accept it as the lesser of two evils. They hope that we can gradually make prosperity world wide, so that as country after country experiences it, each of them in turn will begin to diminish in numbers, and finally we can all be comfortable together in an effortless world. It is conceived as an automatic painless process, occurring naturally and spontaneously and involving no compulsion anywhere. I shall come later to the long-term instability which will prevent such happenings, but there is also an overwhelming short-term reason to prevent it. There is simply no time for it to come about, because everything happens in the wrong order. What is required is the simultaneous existence of high prosperity, social conditions in which the economic disadvantages of parenthood are evident to nearly all classes, and some knowledge of the methods of birth-control. Only under such conditions will the potential parent weigh the rival pleasures to be derived from a motor-car or another child. Such conditions exist in some of the more prosperous countries, and affect the more prosperous members of them, which may be quite a considerable fraction. In others, even among those possessing a high degree of civilization, the fraction affected is very small, and for the remainder there is no such inducement to a careful balancing of the advantages.

Take the case of India. Of some five hundred millions it is doubtful if even one per cent are at present in the appropriate conditions of prosperity, and it is not one per cent but at least sixty per cent that is needed. For the rest the population is increasing at a terrifying rate. As to the prospects of higher prosperity, the risks of local famines are already mitigated by good inter-communications all over the country, and not very much more can be hoped from this. Methods of agriculture can no doubt be improved in many ways, but it would certainly take a long time merely to teach these to such an enormous number of people. And all the time the torrent of procreation continues, itself inevitably decreasing the standards of life. There is, so to speak, simply no time to make the people realize what a pleasant bribe a motor-car is, nor for the matter of that is it likely that enough motor-cars could ever be offered. There are many other parts of the world where the same thing is happening. The colonizations, mainly by the white races, have produced a security of life before unknown, with the immediate consequence of large increases of population, and these increases have automatically lowered the standard of life back towards what it had been previously. The first condition needed for spontaneous limitation itself destroys the chances of establishing the other conditions.

There is still one point to be considered. I have been arguing that there is no chance of spontaneous limitation coming about, because it takes too long, and yet it is even now being actually experienced in some countries. This fact might seem to demolish my argument, but I do not think there is any difficulty in meeting the point. The present era has been unique in that it has combined the wonders of the scientific revolution with the sudden expansion of the white races into vast almost uninhabited regions. The consequence has been that for two or three generations the Malthusian threat did really disappear. In spite of the secure conditions, man could not breed fast enough to catch up with the extending agriculture, and so the other conditions for spontaneous limitation could come into play, before the first one, the condition of security, had killed the chance of them. It would seem unlikely that similar conditions can arise again in world history, so that in estimating future possibilities, there is little prospect that the Malthusian threat will again be overcome spontaneously in this way.

In considering the possibility of the spontaneous limitation of populations, I have been regarding the subject in the manner prevalent, as arising from conditions like those we are experiencing here and now. But it must not be overlooked that there have been many epochs in the past—and perhaps there are cases in the present too—where children were not wanted, and where a more direct solution was found through infanticide. It was usually female infanticide that was practised; this was presumably because the male was more valuable economically as a soldier or workman, but it was also more effective in limiting the increase in later generations. Infanticide is repugnant to all our present systems of thought, and it is hard for us to conceive the state of mind of those who practise it, and so to estimate how it actually works out. It would seem rather likely that it would operate on a lower social level than does our present limitation, because the decision to destroy a new-born child must involve a great emotional crisis, so that it is not likely to be undertaken except in extreme conditions. It will not be the hope of a motor-car, but the pangs of hunger that will bring it about. This form of limitation will hardly come into play in conditions of prosperity, and so it cannot be considered any help in maintaining prosperity.

I have already suggested that the voluntary limitation of populations is an unstable process, whereas any process that is to come about spontaneously has simply got to be a stable process. Here I use the term *stability* in its technical sense, which hardly differs from the popular sense, though it is a little more precise. Stability roughly signifies that, if the system under consideration gets a little above its average level, by that very fact a force comes into play to pull it back, while if it falls below, a force is evoked to raise it again. In this sense the voluntary limitation of populations is evidently an unstable process, but the matter is so important that it may be well to illustrate it by an example. Something of the kind is being already experienced in parts of France, where the peasant population is not maintaining its numbers, but I do not want to be tied to demographic details which are not very accurately known or understood, and the argument is quite strong enough for it to be put in more general terms.

The peasants of province A are not reproducing themselves, with the result that the villages are only partly
inhabited and that part of the land has gone out of
cultivation. Province B on the other hand has an excess
of population, and the land-hunger of the B's will drive
them into taking over the deserted houses, and into
cultivating the neglected land, even though it will have
been the poorest land that had gone out of cultivation.
Some departments of France have in fact already been
partly re-peopled in this way from Italy. Now if the
immigrant B's retain their own customs, they will continue to increase in numbers in their new settlements,
and in a few generations the province A will be fully
populated, but now chiefly by B's. But it may be that
in their new surroundings the B's will feel the influences
which led the A's to decrease, so that they too will start
to decrease, and again villages will be deserted and land
will go out of cultivation. If this happens there will be a
fresh influx of B's, unless perhaps the province B has by
now also got itself into a state where its own population
is decreasing. In that case there will be a new immigration from a province C which has an excess population.
If the C's go the same way after immigration, then D's
will come in, and so the process will go on, with a succession of immigrations, each of which may later fade
out by experiencing the same decrease. But at some stage
one set of immigrants will come in who decline to decrease, and then the province A will experience over-
population. Thus the state of under-population in the
end inevitably cures itself. In a different sense so does the
state of over-population, for the over-population will in-

evitably reduce itself to a condition of exactly full population, either by emigration, or by the starvation of the surplus. The only condition under which the final state of A would not be one of full population would be that there should be no single race on the whole face of the earth that was not stationary or decreasing; if there was a single one that resisted the bribe of the motor-car, that race would people the earth, and this it would do whether its motive was high principle, or some creed, or simply pure stupidity. It is in this sense that I say that the avoidance of the Malthusian threat of over-population is bound to be an unstable process.

I have already shown the short-term difficulties which seem to make it sure that no spontaneous process will avoid the menace of over-population. Is it possible that the statesmen of all countries, perceiving these dangers, should combine together to make and enforce a world-wide policy of limitation? It would have to be world-wide, because if any nation were recalcitrant, its population would increase relatively to the rest, so that sooner or later it would dominate the others. That the prospects of such a world-wide policy are not good is witnessed by the total failure hitherto achieved in the far easier problem of military disarmament. How would the nations settle the respective numbers admissible for their populations? The only principle that would have a chance of acceptance would be to base the numbers on existing populations, and then the question arises why one particular set of proportions between the various countries should be frozen constant for all time. Since the aim of the policy is to retain world-wide prosperity, every single country would be faced with the problem of taking care of its own limitation, and, as has been seen, this would not come about spontaneously. Even if a government could devise an effective method, it would be an odious task for the rulers to have to enforce it, and there can be no doubt they would often evade doing so. With the best of goodwill, it would be hard to enforce the limitation because of the gradualness of the increase, for the rulers could always excuse themselves by the argument that the slight illegal increase of this year was accidental and would next year be compensated by a corresponding decrease, so that action might be postponed, and sometimes it would be postponed too long.

It is clear from all this that the world policy would need to be supported by international sanctions, and the only ultimate sanction must be war. Present methods of warfare would not be nearly murderous enough to reduce populations seriously, and even so they would take a nearly equal toll of victims from the unoffending nations. So after the war the question would arise of how to reduce the excess population of the offending nation. It is not possible to be humane in this, but the most humane method would seem to be infanticide together with the sterilization of a fraction of the adult population. Such sterilization could now be done without the brutal methods practised in the past, but it would certainly be vehemently resisted. I have dwelt on these details, perhaps at unnecessary length, not because I believe they will ever happen, but in order to show that this kind of enforcement, which is the only obvious one, would lead to a condition of strife, jealousy and disorder, which is precisely the condition that it was designed to avoid. The fundamental instability of population numbers cannot be checked by man-made laws, and even if it were successfully done for a few years there is no chance of the system working century after century.

Even worse difficulties, however, would arise than those I have so far contemplated. I have been assuming that the policy of limitation was accepted by the majority on broad rational grounds, but it is quite certain that in a very short time it would encounter fanatical opposition. Even though the procreative instinct has not the violence of the sexual instinct, yet it is an emotion possessed by many people, and as such it will be particularly liable to get incorporated in creeds. There are already creeds that maintain the wrongfulness of birth-control, though there is at present no very strong emotion associated with them. But if there were to be any enforcement of birth-control by authority, it is certain that many new creeds would spring up which would regard the practice as sinful, and the tenet would be held with an enthusiasm not to be overcome by the efforts of rational persuasion. There are many creeds, which we hold to be unwise, which we can admit and leave alone, because their effects are mainly to damage their believers. This could not be one of them, since the believers would automatically gain an undue share of the next generation. Persecution would be the only recourse against such a creed, and the massacre of the innocents or the blood of the martyrs would water the seed of the faith. It is not of course true, as is sometimes maintained by religious devotees, that persecution always fails to extinguish a faith—for example the Arian heresy was much persecuted by the orthodox church, and there are no Arians now—but there is no doubt that persecution is a great encourager, and it is fairly sure that not all such creeds would be extinguished. Once again the effort to produce comfortable prosperity would call for a brutality that is just the kind of thing it is trying to avoid.

It is not only the creation of creeds that may come into play to prevent the artificial limitation of populations; in the very long run a deeper cause will arise to prevent it. Through natural selection animals acquire heritable qualities which fit them to survive, but nature works in a very untidy way to achieve its ends, accepting any method no matter how indirect it may appear to be, so long as it is effective in producing the result. Man has strong sexual instincts, and strong parental instincts, but the procreative instinct, which would make him feel the direct want of children, is much weaker. This did not matter so long as the sexual instinct would ensure the birth of children, but now it is no longer doing so. Nature's untidy method has been defeated by the ingenuity of man. There will be a revenge.

Though the procreative instinct is comparatively weak, it is present in many people, and it is these people who will have larger families than the rest. By the very fact they will hand on the instinct to a greater fraction of the population in the next generation. The process of building a new instinct into the species will certainly be a slower one than the operation of any creed, but it has a permanence possessed by no creed. That an instinct is a very much more powerful thing than any creed may be seen from the sexual instinct; there are many creeds which place the greatest importance on the virtue of chastity, but their prohibitions are seldom effective against the instinct. There is no need for the procreative instinct to become even remotely as strong as the sexual for it to defeat any opposing creed that favours limitation of populations, and so to perpetuate the over-population of the world. Once this stage is reached, nature will have taken its revenge, and there will be little tendency for the instinct to increase further. It is very much of a guess how long such a change will take, but it should be far less than the million years of the change of a species; some analogous considerations which I shall develop in the next chapter suggest it might be something like ten thousand years. After all, for one thing, no very great increase is needed in an already existing instinct, and for another the effect on population from it is so very direct.

To conclude the chapter I return to the narrower question of the tendency of civilization to eliminate its ablest people. This has happened in the past, and is certainly happening now, and if it is always to happen, it signifies a recurrent degeneration of all civilizations, only to be renewed by the incursion of barbarians who have not suffered similarly. If any civilized country could overcome this effect, so that it alone retained both its ability and its civilization, it would certainly become the leading nation of the world. Man is a wild animal, and cannot accomplish this by using the methods of the animal breeder, but may he not be able to devise some-thing that would go beyond the long-drawn-out auto-matic processes of Natural Selection? I think he can. A cruder and simpler method must be used than the animal breeder's. Something might be accomplished on the line of what is called "Unconscious Selection" in the *Origin of Species*.

Unconscious Selection signifies that the farmer, who has no intention whatever of improving his herd, will naturally select his best and not his worst animals to breed from, and in consequence he will find that in fact he does improve the herd. As I have pointed out, we are all the time assessing the rival merits of individuals for promotion; they are each chosen for some special pur-pose, but like the unconscious selection of the farmer, the choice does mark the promoted person as being superior to the average. Any country that could devise a method whereby the promoted were strongly encour-aged to have more children than the rest, would find itself soon excelling in the world. It would only be a rough and ready method, with many defects; for ex-ample, from the point of view of heredity women are as important as men, but it would not so often be easy to take their qualities into account. Furthermore the method would be extremely subject to fashions—in which it would resemble the animal breeder's method—for at one time greatest value would be given to the arts, at another to military skill, and at another to administra-tive ability and so on. However, ability is not usually a very specialized quality, and the effect would be to pre-serve high ability in general, and thereby to increase it, since the abler people would be contributing more, instead of less, than their share to the next generation.

A nation might consciously adopt such a policy, or it might be that an economic policy adopted for quite other reasons should have this unintended result. Whatever way it came about, if it could last for even a few generations, the effect would begin to show. But humanity is capricious and subject to the passions of the immediate present, and it is hardly likely that any country, whether democracy or autocracy, would follow such a policy long enough for it to really tell. The best hope for it to endure would be that it should become attached to a creed, and it would not matter very much whether the creed was reasonable or unreasonable, provided that it produced the effect. Either ancestor-worship, or a belief in the sinfulness of birth-control, would at least place the promoted on an equality with the unpromoted, and with their superior ability this would give them the advantage. But since the matter concerns the more intelligent, a reasonable creed would have a better appeal than a mere superstition. Such a creed might be one which inculcated in those who were promoted the duty of having more children than their fellows, as an act benefiting the human race. The prospect of such a creed arising does not seem very hopeful, but if by its means any country can even partly solve the problem, it will lead the world, and it will be doing so through the method of "Unconscious Selection".

IX

THE PURSUIT OF HAPPINESS

THE present chapter is perhaps a digression from the main theme of this essay, but its subject is such a chief aim of mankind that it seemed to merit discussion. Happiness has been the subject of a great deal of magnificent literature, but it has also been the subject of more trite aphorisms and of more bad philosophy than almost any other in the world. In venturing to discuss it I am acutely conscious that I shall probably be joining the ranks of the bad philosophers, but it has seemed to me that since happiness is of such principal importance to man, I could hardly be justified in evading the subject. My main theme is to be the *pursuit* of happiness, and this is different from happiness itself. A great part of human conduct is dictated by the motive, "I am going to do so and so, because I think it will make me happier," though the prescription is all too often faulty. However that may be, the result is some course of action, and this will affect the external world, so that it becomes relevant to history. It is this that justifies its consideration here.

Before coming to this side of the question, however, it will be well to begin by considering happiness itself. Much happiness comes quite unsought, and in examining their memories for the chief happy or unhappy incidents of their life, most people will find that the really important ones were concerned with entirely intimate matters of a personal kind, which had little relation to the conditions of the external world. Such occasions of happiness or unhappiness will presumably always be among the most important things in life, and since they are independent of the outer world they will continue much the same in the future as in the past. There is undoubtedly a great difference among individuals, in that some are naturally cheerful and others naturally melancholy, and as these are inherent characteristics there is nothing to be done about them, but it is relevant to the present inquiry to ask which type has the greater survival value. To put the matter in its crudest form, is a naturally cheerful person likely to have a larger family than a melancholy one, for if so, then there would be a prospect of a slow increase of cheerfulness throughout the human race. I cannot answer this question at all, but I can see no reason to believe that cheerfulness should triumph, rather than melancholy. There is also the very greatest difference among individuals in the pitch of their emotions, in that some are alternately intensely happy or miserable, while others take both emotions much more placidly. Here again the question arises whether there is survival value in intense emotions, and again it is not easy to answer, though it may be noted that such emotions tend to go with an instability of character which may often lead to forming bad judgments in the other activities of life, and these bad judgments can hardly help in survival.

Among the inherent tendencies of people towards happiness or unhappiness, there is one characteristic, and a very sinister one, which cannot be overlooked. In any boys' school where discipline gets at all slack it is practically universal for there to be bullying. This means that there are many of mankind who positively enjoy making their fellows miserable; it is by no means a majority, but it is certainly not a negligible minority. It is no use arguing that this is only a boyish failing and that in later life the bully will become a virtuous citizen. Conditions in this country give little scope for the exercise of brutality by adults, but this has not always been so, and it is not so in many parts of the world even now: it is the strong arm of the law, and not a change in his nature, that has restrained the bully. It is not easy to see anything that will tend to eliminate him, because his selfishness is a positive help to his survival in all conditions but those of the highest and most stable civilization, and even these conditions only check the expression of his propensities without destroying them. In thinking of the future happiness of mankind, it is a sobering thought that there will be quite a perceptible fraction of humanity that definitely gets satisfaction, and so presumably happiness, from making its fellows unhappy.

Among the more external conditions of human life a great deal of misery is directly due to physical pain, and if these sufferings can be removed either by cure or by means of harmless opiates, it will clearly increase the sum of human happiness. Medical science has already accomplished much in this direction, and it holds promise of a great deal more, so that in this obvious sense man may confidently expect to be happier in the future. But there is another side of physical suffering that it is not so easy to judge about, and that is hunger. If population pressure is to be a main feature of human history, there will usually be a marginal fraction of humanity living on the verge of starvation, which cannot be reckoned as a happy state. However, it is hard to be sure even about this, for the starvation is not usually continuous, but comes in periodically recurring famines, and there is room for happiness in the intervals. To those of us who have never experienced real hunger this may seem unlikely, but it is reported by those who know the Eskimos well, that they are the most cheerful people on earth, and this though they are certainly the race living most continually on the verge of starvation.

In so far as happiness is regarded as an object of pursuit, there is the implication that it is at least partly within the control of the pursuer. Such happiness is a less deep emotion than those I have been considering, and its antithesis can hardly be described as melancholy or misery, but rather as discontent. Much discontent arises from noble motives, but it has regretfully to be admitted that the motive lying behind the widest range of discontent is mere envy, that most unamiable of human characteristics. But whatever the motive, it certainly produces great unhappiness, and it is the kind that stimulates the sufferer into seeking a cure. Man is a very poor prescriber for his own troubles, and he usually sees his grievance, whether real or imaginary, as the only thing in the world that stands between him and a permanent state of perfect bliss. Of course as soon as he has succeeded in removing the grievance, he at once finds another, and this again becomes the most important thing in the world, with eternal happiness once again just round the corner. The target of the pursuit will always evade the pursuer.

It is not recognized by most people that happiness does not come from a state, but from a change of state. That it is so is illustrated by the total failure of every writer to describe a satisfactory paradise, whether in heaven or on earth. The tedium of eternity has almost become a joke, and the descriptions of the earthly utopias are no better. Most of them fail to recognize that the human mind cannot hold any emotion for long at an even intensity, but that it always soon degenerates into something much more tepid. A few authors of course have recognized this. Thus Samuel Butler describes the criticism of the Christian heaven by an Erewhonian, who points out how much better it would be if one always thought that one's wishes were going to be thwarted, and then at the last moment they were fulfilled. Then there is the (almost certainly mythical) American preacher, who told his congregation that heaven would not consist in the playing of harps, but would, like earth, be a centre of busy activity; we can be fairly sure that no "bears" would have been allowed on the celestial stock exchange, but the preacher did recognize that it is a change of state, and not a state, that makes for happiness.

But it is not simply a change of state that makes for happiness; there must be something unexpected about it. Butler's Erewhonian would very soon have got bored by knowing that he was certain not to be really disappointed. Again, in some professions there is an automatic annual increase of salary, but this change is apt to be mentally discounted long in advance; contrast it with the real joy at receiving an unexpected promotion. A great proportion of mankind enjoy gambling. If their conditions are bad, this is easily understandable, because the remote chance of betterment is worth taking; but very many people in secure and prosperous conditions also find it almost necessary to gamble, and this is because it provides just the element of uncertainty, otherwise missing in their lives, that is essential for their happiness.

The external conditions then that are most likely to produce happiness are benefits received at uncertain intervals, and to make the individual continuously conscious of his happiness there must obviously be several such benefits during the course of his life. In the present economic conditions the prescription for a great many people would all too often be ten per cent more pay for ten per cent less work, with the dose necessarily repeated at not infrequent intervals. This is to put the matter very crudely, but it does subscribe to the general human view on the antithesis between work and pleasure. The prescription is of course fantastically impossible of achievement over the course of the ages. Even in a single lifetime the cumulative effect of compound interest would defeat it, and though the son could not expect to start where his father left off, yet he would expect to start above where his father had started, so that the law of compound interest—it is true that it would be at a lower rate—would again come in for the succeeding generations. There is no chance of this sort of thing continuing over a thousand years, let alone a million years, unless there are intervening periods of disaster, to give occasion for a new start. The really wonderful thing about the last century has been that exciting improvements of condition have been happening at frequent intervals for about six generations. And even so, it is not very evident that those living in the present conditions of enhanced prosperity are any *happier* than the people described by Dickens.

A chief question, from the point of view of this essay, is whether there is any survival value in happiness. Are the naturally happy people more likely to be the ancestors of future generations, than are the rest, for if they are, then a greater number of the future race will tend to inherit this happy disposition. The answer is very doubtful, and it may well be negative. The reason lies in the fact that contentment is not a stimulus to action like discontent. It must of course be recognized that there is a good deal of what I may call stimulated discontent, for many political leaders find it useful to stir up discontent among their adherents, even though these may really be of the contented type. Leaving aside this stimulated discontent, a man, who has the spur of his own genuine discontent to drive him, will struggle harder to achieve success than will the contented type. On the average, he will be more successful, but the success will not content him, so that he will always be spurred on to further efforts. If this success is, as in the long run it will be, associated with his making a greater contribution to later generations, it follows that the discontented type will increase in numbers at the expense of the contented type. This argument leads to the disappointing conclusion that future man will be more discontented than man of the present day. I do not want to press it strongly but in the light of it, no matter what the future conditions of life may be, there seems absolutely no reason to expect any notable increase in the sum of human happiness.

In connection with the matter of human happiness it is a very pertinent question to ask whether man really enjoys being civilized, for on the answer to some extent depends the stability of future civilization. In the past there have been so many cases of the decay of civilizations, that it is rather tempting to believe that the majority really find a state of barbarism more congenial. Thus the civilization of the Mayas had already seriously decayed under the rule of the Aztecs, long before it was destroyed by the Spaniards. Again the Roman Empire was destroyed by the onslaught of the Germans, in spite of the fact that it had been steadily, and on the whole successfully, civilizing many of them for two or three centuries before the collapse; they found the barbaric life more satisfying. To take a modern example, the Republic of Liberia was re-peopled by negroes returned to it from America. These had seen civilization, even if they may not themselves have gained much profit from it, but anyhow they showed little wish to avoid the relapse.

There are no doubt many causes that have led to such relapses into barbarism, but a chief one is the existence of the class of men I have called bullies. Such men are apt to be brave and self-confident, but selfish and concerned only with their personal interests, and above all indifferent to the sufferings of those around them. Such men, always ready to assume leadership, only interested in their own advantage, and indifferent to the fate of their fellows, are perfectly adapted instruments for destroying the delicate balance of civilization.

Now though it is indisputable that many civilizations have relapsed into barbarism, each of them must after all have grown out of barbarism before it could relapse, so that instead of arguing that man has relapsed into barbarism on account of his dislike of civilization, one might argue with almost equal force that he has become civilized because he does not like barbarism. The best answer to the question which of the two he prefers can be given by examining the parallel of another human taste. Originally man was a hunter, and very many people still retain the trace of it in that they find a spontaneous, almost instinctive, joy in the chase, which they can get in no other way. About ten thousand years ago there came the agricultural revolution. This was a totally new thing for man—and indeed for the whole animal kingdom, except for the independent discovery of it by a few insects. At first it can have had no emotional appeal at all, but rather the reverse; the discoverers must have felt they were doing a disagreeable and tedious job and they only did it because it was so clearly advantageous. For a long time the great majority would retain the emotions of the hunter, and would only cultivate the soil under the impulsion of stark necessity.

The practice of agriculture was an *acquired character*, and so it had to be acquired anew in each generation, and many must have revolted against the tedium and gone back to the more congenial practice of hunting. But there would be some whose nature was more tolerant of farming who would stay farmers, and those of their sons who inherited the taste would continue on the farm, while their brothers would drift back to hunting, so that there would be an "unconscious selection" towards agriculture. The new habit of life would gradually establish itself in some men's heredity, carrying with it an emotional appeal which might ultimately become as strong as had always been the appeal of hunting to the rest of the race. I do not know if biological principles could tell how long the creation of such an instinct would take, but there is no need to ask the question, because we have the answer before us. There are a great many people now existing—that is, certainly after less than ten thousand years—who undoubtedly have the instinct for agriculture; these are the people who derive deep emotional satisfaction from gardening, even when they are in no way driven to it by economic necessity. Though Voltaire might be claimed as a typical product of the age of civilization, he really belongs to the age of agriculture, for he represents Candide, in his disgust at the world round him, as finding his ultimate satisfaction in the cultivation of his garden.

The same sort of thing must be happening with the urban revolution. In China and the Levant this goes back for several thousand years, and already the promotion of the taste for civilization from an acquired to an inherent human character must have begun, but for most of the world this can hardly be so yet. After all, in western Europe and America, it is hardly more than thirty generations since most of the ancestors of the present city dwellers were completely barbarous, and it is therefore not surprising that many of their descendants should regard civilization as a disagreeable necessity without emotional appeal. This explains why so many civilizations have been rather short-lived, but it also answers the question whether man really likes being civilized. There can be no doubt whatever of the advantage of being civilized, in that it permits of larger populations, and these will prevail by their numbers against the smaller populations of barbarism. Some of the citizens may not like being civilized at the present time, but that does not matter, for in due course their descendants will grow to like it emotionally and instinctively, by the same process of "unconscious selection" from among them, as has happened with agriculture. The process has begun already, and in the course of a few thousand years at most, a great fraction of mankind will feel spontaneously the emotional appeal of civilization.

Let me follow this train of thought to its conclusion. If the agricultural revolution has followed hunting into our instincts, and if the urban revolution is going the same way, what about the scientific revolution, which has only just begun? The majority of mankind certainly have no taste for science; they regard the subject as a disagreeable necessity only practised for its obvious material advantages, and they relapse from it with enthusiasm towards the more instinctive tastes of the earlier revolutions. Nevertheless, the advantages of the new acquired character are evident, and there can be little doubt that it will follow the same course as the previous ones by the gradual selection of those who find the new system naturally congenial. In this way I shall expect that before the end of ten thousand years, science will make an emotional appeal to the instincts of a majority of the human race, of the same intensity as the emotions they now derive from the arts of the city, from the garden and from the chase.

X

THE HISTORY

A HISTORY of the future is different from a history of the past, because it cannot in any sense be a narrative. It cannot say what will happen in anything like the same manner as past history says what did happen. All it can do is to say what things will be happening most of the time and in most places, but without being able to specify those times and those places. This it does through consideration of the laws of nature, chief among which is the law of human nature. In the preceding chapters I have given what seem to me to be the main principles of this law, and if I have given them correctly or even roughly so, it will be seen that the general trend of history is inevitable. It becomes hardly more than a summary of my previous discussions.

The variety of happenings of a million years is obviously so prodigious that, at some place and at some time, almost anything that could be thought of will be found to have occurred, and so a prophet can foretell what he likes with the fair certainty that an example of it could be cited before the end of the period. I should not be content with such a verification of my predictions. I want to foretell the things that will be happening most of the time and over most of the earth, and I should count it as defeat if the historian of a million years hence should point out that my forecasts were verified, because they did once happen for a few decades on some remote island in the Pacific. Indeed I might express my ambition better by putting it the other way round. At the end of a million years some Gibbon, with all the vast archives of the world at his disposal, may undertake the stupendous task of writing the whole history of the human race. In the excessively unlikely event of his reading this work, I should be best content if he considered it as unworthy of mention, because it was a mere description of all the things that were entirely familiar and therefore uninteresting, the things that all his readers would take for granted. He would feel free to pass them over, and spend his time in describing the more exceptional and remarkable things that had happened in the course of the ages.

Before coming to the details it may be well to remind the reader once again of the operation of the law of large numbers in connection with probabilities. In the events of the world one cannot of course actually give numerical values to the odds as one can in a game of chance, but I can use the analogy to show what I mean. If I said that the odds were two to one on such and such a state of the world as compared to some rival state, I should not mean that it was twice as likely that the favoured state would be happening all the time; I should mean that in the course of the ages it would prevail for about two-thirds of the time, and the rival state for one-third. Now there can be no doubt that most things in the world fall under the category of large numbers— the mere fact that there are even at the present time two thousand million individuals guarantees this—so that probabilities become certainties in the sense that very probable things will be happening most of the time, while less probable things will still happen, but only for a small part of the time. But there may be occurrences so rare that the law of large numbers cannot be applied to them at all; for example the discovery of the New World in the fifteenth century was a unique thing, because there were no other new worlds to discover. Or again there is the unlikely, but possible, chance that there should be a collision of the solar system with another star, which would destroy all life on earth. If any such rare event should occur, it would upset all predictions, and there is nothing more to be said about it.

There are no doubt readers who will dislike many of the things I am forecasting and who will try to evade them by the hope that one of these rare unforeseeable chances will entirely alter things, and lead to a condition of the world more to their liking. It is possible, but it is much more likely that such things will be unfavourable than favourable. Whereas small changes produced by chance are as likely to be beneficial as detrimental, when it comes to large changes, the probability is that they will be unfavourable. I have already cited an example of this from the science of genetics, where, by means of X-rays, changes can be induced in the genes of the cells of animals. If the change is small, it may benefit the animal, but if it is large it is almost invariably deleterious, and often lethal. The balance of the natural forces in an animal is so delicate, that any large change in one feature upsets it entirely; only if there were compensating large changes in other features could the condition of the animal be improved, and there is practically no chance of these other changes happening to occur simultaneously. A similar principle must apply to the delicate balance of interactions which go to make up the life of the human race. Thus anyone who hopes that some rare, large, unforeseeable occurrence may better the fate of humanity is almost certain to be disappointed, for it is enormously more likely to worsen it. The best hopes of benefiting humanity are to be based not on this, but on the working of small changes and the law of large numbers, by which there is at least some prospect little by little of improving the condition of the world.

In what follows I shall divide up the principal activities of humanity under the headings of population, economics and so on, and consider each briefly in turn. It may be well to repeat that the views I put forward on these subjects are not intended to be exclusive. It is to be expected that there will be many happenings that contradict them; I am only claiming that such happenings are likely to occur a good deal less frequently than the conditions described here.

POPULATION

The central feature of human history must always be the pressure of population. Man, the wild animal, will obey the law of life and will tend to multiply until he is limited by the means of subsistence. This is the normal condition of the world, and it carries the consequence that the final check on population is by starvation. There will be a fraction of humanity, a *starving margin*, who have got to die simply because not enough food can be grown to keep them alive. The death may be directly due to intermittent famines, or to diseases caused by malnutrition, or it may be due to warfare; for when a country is dying of starvation and sees, or thinks it sees, a neighbouring country with plenty to eat, it would be beyond most human nature to accept certain passive death instead of possible active death. The central question for humanity is the problem of the starving margin.

To those of us living the life of Europe at the present time this is a shocking fact, implying a condition so unfamiliar that there are many who may not willingly believe it. This is because of the quite exceptional history of the nineteenth century, during which, in spite of enormous increases of population, many countries had no starving margin at all. The disbelief may be helped by the fact that the population of some countries has recently started to decrease. Such decreases have occasionally happened before too, but, as I have argued earlier, they constitute an entirely unstable state of affairs, in that the nations which are decreasing in numbers will die out, and will be replaced by the starving margins of the others.

On the time-scale I am considering, the action of starvation can be treated as if it were uniform and continuous, but it is fortunate that it would not appear so to the individual, for famines are not like that. Since man can never aspire to the real control of climate, there will always be fluctuations in the harvests he can produce. For some years there may be a sequence of good harvests, and starvation will be forgotten by everyone, but after that a few bad harvests will fatally redress the balance. So it would be wrong to imagine that the starving margin suffers a life of continuous grinding misery, but rather one of misery alternating with a precarious prosperity. Even so there are many at the present time who will regard this state of affairs as very dreadful, but, as I have already pointed out, it has always been the normal condition of life of the Eskimos, who have the reputation of being the most cheerful race on earth. So, as far as concerns the individual, the starving margin would not be in a state of continuous misery, but rather of misery alternating with happiness, which after all is not very far below the state of the rest of the world. For history regarded on the long-term scale, however, these fluctuations of prosperity disappear, and the fact has to be faced that it will be starvation that limits the numbers of the human race.

The effects of over-population will be a chief feature to be considered in the later sections of this chapter, but here the question arises of what the total population of the world is likely to be, and the answer is immediate. Whatever food the efforts of mankind may produce, there will always be exactly the right number of people to eat it. It all comes back to Malthus's doctrine and to the fact that an arithmetical progression cannot fight against a geometrical progression. If at any time some discovery, usually an agricultural one, should make a greater supply of food available, then, reckoning on the long-term time-scale, instantaneously the population will rise to the new level, and after that things will go on as before, but now with a larger starving margin in the larger population. It is by no means evident that the world will be any the better for it, but the point is not whether it is a good thing, but whether it will happen, and the answer is that undoubtedly it will. The social sense of any community, and its immediate practical interest, will not tolerate living in contact with the sufferings of its own starving margin, if it is in any way possible to relieve them. The relief will all too frequently involve bad agricultural practice which will ruin the land in the long run, but short-term necessity will always prevail against long-term prudence. What is the good of telling a man that he must die now for fear that his grandson may be short of food a century hence? So all over the world there will be immediate pressure to produce more food, and the forecast of the future numbers of mankind is the same thing as the forecast of the future of agriculture, but unfortunately it will all too often not be the ideally best agriculture.

I do not know how far it would be possible at the present time for an agricultural expert to forecast the total amount of food the earth could produce, but I am certainly in no way qualified to do so myself. I shall therefore, though only very tentatively, set down a few considerations on the subject. In the state of wild nature animals and plants have learnt to live even in the most unfavourable sites, which they have been driven to occupy through the intense pressure of natural selection. This suggests that the total amount of living matter of all kinds on earth can never be very different from what it is now. It is true that new ice ages or pluvial periods, which we cannot foretell, might bring rain and therefore fertility to the present deserts of the earth, but even if there were no compensatory loss of fertility elsewhere, this would hardly even double the area available for life. So it may be assumed that the total living matter of the earth is roughly constant, and all that man can hope to do is to convert more of it to his own use. This he does by promoting the growth of particular types of plant at the expense of the rest; it does not increase the total amount of living matter, for there must be less vegetable life in a wheat field, than in the same field when it is let to run wild. Now under the pressure of his needs man has already exploited to a very great extent the more fertile soils in many parts of the world, but he has only succeeded in replacing the wild plants by food plants through the liberal use of fertilizers. There are still no doubt a good many parts of the earth where this has not yet happened; in particular this is true of the New World where the pressure of population has not yet become at all severe. But on the whole to develop further food supplies means devoting inferior lands to agriculture, and such lands will call for an even greater use of fertilizers. So the possibility of greater supplies of food may be assessed by the available supply of fertilizers.

It may then be that the future numbers of humanity will depend on the abundance in the surface of the earth of the chemical elements which are necessary for life. Most of them are abundant enough to raise no difficulty, either because they occur in practically unlimited quantities, or because only small quantities are needed. Two only deserve comment, nitrogen and phosphorus. The supply of nitrogen in the air is quite unlimited, but it is not easily available to plants by natural processes, and to supply it in sufficient quantities for agriculture demands a considerable amount of mechanical power. This method of getting nitrogen is of course already common practice, and provided enough work is done to win it, there seems no reason to think that nitrogen need ever run short. The question of phosphorus is far more serious, though less of it is needed. At the present time it cannot be said to be actually in short supply, though even now it is commercially very profitable to mine fossilized phosphorus deposits, and they are used even in the soils which are naturally fertile. There are great tracts of land, in particular in Africa, which are permanently deficient in phosphorus, and these can never be raised to the fertility of the more favoured regions, unless large quantities of it can be supplied to them. So it may well be that the future numbers of the human race will depend on the abundance of phosphorus in the earth's surface.

I have so far only considered extensions of the methods of ordinary agriculture as the way to increase food supplies, but there remains the possibility that wholly new methods might be discovered. All existing animals depend on the vegetable kingdom for the supply of the constituents of their bodies, but man might aspire to free himself from this limitation. It may well be that some day it will be found possible to synthesize from their component elements some of the exceedingly complicated molecules which make up the important proteins. The essential first step is to do this on the laboratory scale, but even if this was accomplished it would be a very different thing to make them in bulk, and it would constitute a problem of chemical engineering very far beyond any that has yet been dreamed of. It is perfectly open to anyone to disagree, but I simply cannot believe that there will ever exist factories capable of turning inorganic materials directly into food, so that they should be able to do it on a scale which could supply the diet of thousands of millions of mankind. Unless it could be done on this scale it would not have any material effect on the numbers of humanity.

There remains the possibility that new types of vegetable should be converted into food fit for man. I have already touched on the possibility that man might some day make grass into an article of human diet, which is in effect only to say that he might discover a more efficient way of eating it than through the medium of beef. But it is to be remembered that the ox has to graze most of the time in order to get enough protein even for its own body, and this shows that only a small fraction of the grass could be really useful to man. The process of directly extracting the protein might be more efficient than making the ox do it, but it would hardly be hundreds of times more efficient. And it is at least possible that, when the plant-breeder had modified the grass into being rich in proteins, he would find it demanded fertilizers on such a large scale, that it would be more profitable to use them instead for growing wheat.

A quite different suggestion that has been made, is that food supplies could be increased to an enormous extent by the cultivation of the vast areas of the ocean. The prospects do not look at all good. We know that every spring the plankton grows so fast that in a few weeks it has stripped the upper layers of the ocean bare of some of the chemical salts needed for life. To get large food supplies out of the sea would therefore demand much more than the mere harvesting of the plankton, though this would itself be a very formidable task indeed. Either it would be necessary to expend an enormous amount of power in churning up the ocean, so as to make available the salts from the unimpoverished depths, or else fertilizing chemicals would have to be poured into the sea on a quite fantastic scale.

I shall not pursue such conjectures further, since, when unmade discoveries are admitted to be possible, the subject becomes so uncertain that it is hardly a profitable field for close argument. Nevertheless, I shall risk saying what appears to be the most probable forecast of the future numbers of mankind, though I need not say, I recognize that it may be completely upset by some unforeseen discovery. In view of the fact that it is only the existing vegetable kingdom that can be exploited, I do not believe there will be any revolutionary changes in agriculture but only steady improvements; the improvements will, so to speak, be described by increases in percentages, not by multiples of the present yields. The world will be covered by a population of the same sort of density as is now found in its richer agricultural districts, in countries such as China, India or much of Europe; but, in reckoning this, allowance must be made for differences of climate and of the natural fertility of the soils. In effect this will mean no great increase in the populations of Europe and Asia. The soils of Africa are for the most part not so good, but there is room for some increase there. There should be great increases in the Americas, and considerable ones in Australia and in some of the large Pacific islands. As I have pointed out short-term necessity is often likely to interfere with really good cultivation, but even if this good cultivation could be assumed, it may be estimated that the population of the world is never likely to be more than about three to five times its present numbers.

GOLDEN AGES

The conditions of population pressure must be expected to be the world's normal state, but it is not of course a constant state, for there have at intervals been what may be called *golden ages*, periods when for a time a part of the world could forget about the starving margin. There has tended to be a certain warping in the proportions of history, as given to us by historians, perhaps because it has been chiefly during golden ages that there has been sufficient leisure for anyone to become an historian. At all events the great histories of the world have been written in such periods; Herodotus, the father of history, wrote during the commercial boom of Athens, Tacitus in the great days of imperial Rome, Gibbon at the height of the eighteenth century Age of Reason, and however much they were depicting less favourable times, their views were inevitably coloured by the conditions that they saw round them. Now we are living in or perhaps at the end of a golden age, which may well prove to have been the greatest golden age of all time, and we too are apt to be warped by the feeling that it is a normal time.

Many readers may be shocked at first at the thought that the past century, an epoch so often decried for its many faults, should have been the greatest of golden ages, but I think it can be justified. In past golden ages the prosperity was usually at the expense of other peoples; for example, Rome prospered by looting the east and enslaving the barbarians of the west. Our golden age came about with comparatively little harm to others; it was mainly through mechanical discoveries which made possible transportation on a great scale, so that vast new areas of the world could be opened up for agriculture. It is true that this was done largely at the expense of the American Indian, and his treatment often does not make a pretty story, but still it was a case of many hundred millions prospering at the expense of a few millions, and so the proportion of suffering inflicted to benefit received must have been far smaller than in most of the previous golden ages. The chief benefit was of course to the white races of the Atlantic seaboard, who for more than a century have been able to forget about their starving margin, but it has by no means been limited to them, for many of the other races have benefited too, as is witnessed by the great increases of population of India and Africa, though in these parts of the world they have not been so easily able to forget their starving margins. We are again becoming very conscious of the world's population problem, but now there are no frontiers or unknown parts of the world into which to expand, and so our golden age is probably near its end.

In the future there will of course be other golden ages, but it can hardly be expected that the balance between good and ill will often be as favourable as it has been in the recent one. It might be that, either by conquest or by commercial exploitation, some region should gain mastery over other regions, to such an extent that it could relieve the starvation of its own margin at their expense. The conquering nation would flourish and call it a golden age, forgetting that its prosperity was at the expense of the peoples it had overcome; it would be very unlike the colonial exploitations of our own age, which, even if they are open to criticism in some ways, have in most cases increased the populations of the colonies. Another possibility that might create a new golden age is that some discovery should make available a vast new source of food, and that consequently there would be enough food for perhaps double the previous population of the world. At once there would be a golden age, but after a very few generations the result would be even more desperate than before, for there would be a starving margin of people now twice as great. This in effect is not unlike what has been happening recently, but the present age has had an advantage, never likely to be repeated, in that it started at a time when the civilized world had frontiers over which it could expand, and now it has abolished all frontiers by expanding over the whole earth. Unless there should be a catastrophe to the world beyond all thinking it can never contract to such an extent that there would again be frontiers, and it is only if this happened that it could have the chance of again exploiting the vacant places of the earth, so that only under these conditions could there be another golden age, which in any sense would match the present one.

SCIENCE

It is the fashion at the present time in some circles to decry the value of scientific discovery, and to claim that it is responsible for all our ills—no doubt there was a similar fashion ten thousand years ago to decry agriculture. This view simply will not bear examination. No one would dispute that there are some new troubles in the world which were not foreseen, but they have come about precisely through the successful solution of problems which man has always been trying to solve, though never before with much success. He has always aimed at making a better life by curing disease, by prolonging life and by enlarging his communities so as better to spread and share the risks of the world. Suddenly through the methods of science, in particular by the new methods of communication and transportation and by medical science, he finds that all these aims are achieved, but he discovers that they lead to new troubles he had not had the imagination to think of. So now he is blaming those who have done exactly what he asked, because he finds he does not like a few of the consequences, and he forgets that he is all the time receiving benefits out of all proportion to these troubles.

The benefits of science which affect the ordinary man directly are due to such things as medical science and the transportation of foodstuffs, together with things like the electric light and the telephone or radio, which might be classed rather as luxuries than necessities. These would never have arisen but for the developments of pure science, which is primarily an intellectual pursuit, studied for its own interest rather than for any intention of benefiting humanity. It is fortunate that there are many men who are driven by this purely intellectual urge, for knowledge would never have advanced far if it had only been stimulated by the motive of practical benefit to humanity. It is the pure scientist who has opened up new realms of thought to the rest of the world, and the advance continues.

There seems to be no bound to the field of scientific thought, but nevertheless in an opposite sense every new discovery does set a bound by excluding alternatives which had before been regarded as admissible. In this second sense the field narrows; for example, it is not permitted now to doubt the validity of the laws of thermo-dynamics—laws which were quite unknown little more than a century ago. But this is not the occasion for a technical discussion on the future of the physical sciences and I will only say that, whatever new ideas may come up—and there is every sign that there will be many of them—there is still plenty of room for improvement inside the known fields. In the hard times to come it is not to be expected that the remoter speculations of pure science will be pursued as energetically as its practical applications; for example, metallurgy and chemistry will appear more important than astronomy, and fortunately there seem great possibilities for development for a long time in these practical sciences. Nevertheless we may confidently expect that there will be some who, like Faraday, still hear the call of pure science, for it is from them that the really great advances will originate.

It is in the biological sciences that the most exciting possibilities suggest themselves, perhaps because biology has only recently shown rapid advances like those made earlier in the inorganic sciences. I will only speculate on a few among these possibilities which might have great effects on human life. I have already referred to the possibility of quite new sources of food, and I need not enlarge on that further. Another type of discovery may be connected with hormones, those internal chemical secretions which so largely regulate the operations of the human body. The artificial use of hormones has already been shown to have profound effects on the behaviour of animals, and it seems quite possible that hormones, or perhaps drugs, might have similar effects on man. For example, there might be a drug, which, without other harmful effects, removed the urgency of sexual desire, and so reproduced in humanity the status of workers in a beehive. Or there might be another drug that produced a permanent state of contentment in the recipient—after all alcohol does something like this already, though it has other disadvantages and is only temporary in its effects. A dictator would certainly welcome the compulsory administration of the "contentment drug" to his subjects.

Another possible, though rather remoter, discovery suggests the most curious consequences; this is the control of the relative numbers of the two sexes. It is known that the sex of a child is carried by the sperm, not the ovum, and it is at least imaginable that some method could be found for sorting out those of the sperm cells which carry the male or the female character. It would thus become possible to regulate how many men or women there should be in a population. If such a practice could be developed it is sure that for a time there would be a great unbalance in populations. A nation with ambitions for conquest would produce a large number of men for its soldiers, but would pay for it by not having enough women to give birth to the soldiers for succeeding generations. On the other hand, just as the stockbreeder keeps few bulls and many cows, another nation might decide that it needed few men in order to maintain its numbers. Would such a predominantly female population be able to stand up against the male one, or would a "rape of the Sabine women" rectify the disproportions?

It is clear that the most remarkable effects will be produced if such developments in biological science should come about, and it is impossible to conjecture how they will turn out. I can only record the opinion that in the long run their effects will mainly cancel out for the reason I have developed in an earlier chapter, that man is and will continue to be a wild animal. To produce effects of these kinds there must be a master, and the master must be above and not subject to the procedure he is enforcing on his subjects. The dictator could not afford himself to take the "contentment drug", because if he did so his capacity for rule would certainly degenerate. It always comes back to the same point, that to carry out any policy systematically in such a way as permanently to influence the human race, there would have to be a master breed of humanity, not itself exposed to the conditions it is inducing in the rest. The master breed, being wild animals, would be subject to all the fashions, tastes and passions of humanity as we know it, and so would never have the constancy to establish for generation after generation a consistent policy which could materially alter the nature of mankind.

In connection with the recent wonderful advances in medical science, this is the place to mention a matter that will very soon indeed be of immediate importance. Since in the normal condition of the world there will be a margin of every population on the verge of starvation, it seems likely that there will have to be a revision of the doctrine of the sanctity of the individual human life. In the old days the doctors were under the obligation of doing all they could to preserve any life, though they had no great success in their efforts; now it is hardly too much to say that most diseases have come under control, or anyhow to judge by recent progress most of them soon will. But is the world the better for having a large number of healthy people dying of starvation, rather than letting them die of malaria? One of the justified boasts of recent times has been the great decrease that medicine has made in infant mortality. Whereas in the old days a mother might bear ten children and have only two survive, now she may bear only three and she will be regarded as very unlucky if all do not survive. But the difficulty in the world is going to be that the

number of people born is too great for the food supplies, so that a fraction must die anyhow; may it not be
better that they should die in infancy? The truth is that
all our present codes about the sanctity of human life are
based on the security of life as it is at present, and once
that is gone they will inevitably be revised, and the revision will probably shock most of our present opinions.

ECONOMICS

A very great change in world economics is inevitable
when the accumulated stocks of coal and oil are exhausted. In the scale of human lives this will of course
be a gradual process, marked by their slowly growing
rarer, but on the scale of a million years the crisis is
practically with us already. We shall have spent the
capital accumulations of hundreds of millions of years,
and after that we shall have to live on our income.
Everything depends on whether a substitute can be
found which provides power out of income at anything
like the rate at which we are now getting it out of
capital.

In an earlier chapter I have reviewed the possible
sources of energy, with the conclusion that none are
going to yield it up easily. The energy is there in sufficient quantity, but it will take an enormous organization to get it into usable form. A very much greater
fraction of mankind will be needed to mind the
machines, than are at present needed to get the coal out
of the mines. And there is another difficulty which may
arise. If it should prove impracticable to get the energy
directly from sunlight, there is the possibility of getting

it by the intensive growing of vegetables, say by turning potatoes into industrial alcohol. But if there is always to be a margin of starving humanity, is it not probable that the potatoes will all have to be eaten before ever they are allowed to reach the distilleries? A necessary condition then for getting energy out of vegetables is that it should be found possible to grow the vegetables under conditions where they do not require soil that might be used directly for food production.

To provide energy on the sort of scale to which we are accustomed will call for a very elaborate organization, a great many machines, and a great many people to mind those machines. In view of the shortsightedness and unreliability of human nature, it seems rather unlikely that any process of this kind could be made to work on a world-wide scale for century after century. But it does seem very possible that some part of the plan should be carried out, so that there should be a considerable supplementation to the large amount of energy we already get from water power, which does of course provide energy out of income.

The general picture of the economic condition of the world then is that the chief centres of power production, and so of the most elaborate civilization, will be the regions where there is water power, that is speaking rather loosely, mountainous regions. It will be these that are the centres of manufacture, and they will exchange their manufactures for the surplus food produced in the agricultural regions. There will also be large "power farms" in various parts of the world, storing energy, either by some direct mechanism, or through the intermediary of vegetables. It may be guessed that it will be what I may call the mountaineers, who possess the most readily available energy, who will become dominant; through their wealth they will tend to have the highest culture, since culture most easily comes from the leisure created by wealth. It will be they who will tend to rule the world on account of their economic advantages, and to judge by most past experience they will be hated by the others for it.

There will be the same sort of contest of interests between the mountaineers and the plain-dwelling agriculturists, as there is even now between town and country. Most of the time the mountaineer will have the advantage, but the farmer being the food producer is bound to have the advantage in times of famine, which will not be infrequent. And there will be parts of the world that relapse frankly into barbarism; they will be the less fertile regions which could not produce much food, so that the more civilized people would get no advantage from exploiting them. But there will be other regions which also relapse into barbarism, though the fertility of the soil could support a greater population than it in fact bears. It is to be expected that such a state of affairs will not usually be tolerated by the civilized countries, who will conquer them, and export their own starving margins to fill up the vacant places.

I can make no claim at all to anything but the most superficial knowledge of the highly technical subject of pure economics, and the following speculations must be read in the light of this defective knowledge. In the economics of exchange I will not conjecture what sort of medium will be used. The metallurgical value of gold is not very great, and its mystical value is dead, so that it is not to be expected that it will survive. It would seem that in the long run there is likely to be some *uncontrollable* medium of value, functioning in the same sort of manner that gold used to do, instead of the present manipulated systems which are so liable to political abuse. Though no doubt there will be variations through the ages, it is hard to think of anything having a greater simplicity than a monetary system and therefore presumably that will prevail most of the time. It is quite safe to say that there will always be rich and poor. Wealth will be the mark of success, and so the abler people will tend to be found among the wealthy, but there will always be many among them of a far less estimable character. These are the people who are interested not in the work, but only in the reward, and they will all too often succeed in gaining it in a variety of discreditable ways, such as by currying favour with an autocrat.

As to the less successful members, the standard of living of any community living on its real earnings, as the communities of the future will have to do, is inevitably lower than that of one rapidly spending the savings of hundreds of millions of years as we are doing now. There will also be the frequent threat of starvation, which will operate against the least efficient members of every community with special force, so that it may be expected that the conditions of their work will be much more severe than at present. Even now we see that a low standard of living in one country has the advantage in competing against a high standard in another. If there is work to be done, and, of two men of equal quality, one is willing to do it for less pay than the other, in the long run it will be he who gets the work to do. Those who find the bad conditions supportable will be willing to work harder and for less reward; in a broad sense of the term they are more efficient than the others, because they get more done for less pay. There are of course many exceptions, for real skill will get its reward, but in the long run it is inevitable that the lower types of labour will have an exceedingly precarious life. One of the triumphs of our own golden age has been that slavery has been abolished over a great part of the earth. It is difficult to see how this condition can be maintained in the hard world of the future with its starving margins, and it is to be feared that all too often a fraction of humanity will have to live in a state which, whatever it may be called, will be indistinguishable from slavery.

POLITICS

In the political sphere it must be recognized that there have always been a great many different forms of government which have shown that they can work in practice, and so it is to be expected that the same will be true in the future. The world will be a sort of museum of different methods of ruling mankind. There will be autocracies, oligarchies, bureaucracies, democracies, theocracies and even peaceful anarchies, and no doubt each of them will produce a special political philosophy intended to justify its own procedure against all rivals. In such a variety it is not possible to foresee any detail, and I shall only touch on a few generalities.

Whatever forms the government may take, there can be little doubt that the world will spontaneously divide itself into what I shall call provinces, that is to say regions, though with no permanently fixed boundaries, which possess some homogeneity of climate, character and interests. I use the same word whether the different provinces are federated together, or whether they are what we should now call separate sovereign states. How large will these provinces tend to be? That will depend on the means of communication and transport, and so once again there arises the question of whether the fuel problem is solved wholly or partially or not at all. In the past the chief means of communication was the horse, and the countries of Europe are still mostly of a size adapted to suit this almost extinct means of transport, though some of the more newly formed ones do show a trace of the influence of the railway. None of them are really of a size suited to the motor-car or the aeroplane, or to present power production, whether by coal or water-power, which cuts right across the national boundaries.

If the fuel problem is solved completely, so that mechanical power and transportation is available in the future to a greater extent even than at present, then the provinces will be large; for example, the whole of Europe may well be one, and the whole of North America another. Even if no solution were found to the fuel problem, the world would not revert to its old conditions, because, even if transportation became difficult, intercommunication would still be easy by telegraph. The horse might become important again, and at a guess the provinces would tend to be about as large as the present countries of Europe. In this case, with its greatly increased future population, North America might be expected to break up into a dozen provinces or so. My own conjecture about fuel has been that something intermediate will happen, in that power supplies will not be as easy as they are now, but that by greatly increased effort they will be brought to something not very far below the present level. In this case North America might fall into four or five provinces, and western Europe into one or two; it is never to be expected that there will be any permanence about the numbers of them or about their boundaries.

Consider next what are likely to be usual relations between the provinces. It is too much to expect that there can ever be a permanent world government benevolently treating all of them on a perfect equality; such an institution could only work during the rare occasions of a world-wide golden age. To think of it as possible at other times is a misunderstanding of the function of government in any practical sense of the term. If the only things that a government was required to do were what everybody, or nearly everybody, wanted, there would be no need for the government to exist at all, because the things would be done anyhow; this would be the impracticable ideal of the anarchist. But if there are to be starving margins of population in most parts of the world, mere benevolence cannot suffice. There would inevitably be ill feeling and jealousy between the provinces, with each believing that it was not getting its fair share of the good things, and in fact, it would be like the state of affairs with which we are all too familiar. If then there is ever to be a world government, it will have to function as governments do now, in the sense that it will have to coerce a minority—and indeed it may often be a majority—into doing things they do not want to do.

In the light of these considerations it is to be expected that a single government of the earth will not arise very frequently. Most of the time the provinces will be nearly independent states, which form alliances with one another so as to compete against rival alliances. It will be the old story of power politics again. Now and then a Napoleon may arise, and unite some of the stronger provinces, and with their help he may overcome the rest. For a time he will form an unquiet world government, but after a time his dynasty will decay and the world will go back to the condition of the contending provinces. Here again much depends on the fuel problem. If transportation is easy, world conquest will be easier both for military reasons and because the more uniform culture should make the world government more acceptable.

For the government of the separate provinces it is no use hoping that democracy could often be possible, for the very simple reason that a hungry man will vote for his next meal, rather than for reasons of state. Even at the present time the attempt to import democratic institutions into poverty-stricken countries has been a failure. A necessary condition for democracy is wealth, and the wealth must not be concentrated in too few hands; the lack of this diffusion of wealth is the reason why some rich countries, such as imperial Rome, failed to give democracy to their peoples. Widespread wealth can never be common in an overcrowded world, and so in most countries of the future the government will inevitably be autocratic or oligarchic; some will give good government and some bad, and the goodness or badness will depend much more on the personal merits of the rulers than it does in a more democratic country. Occasionally through conquest, or perhaps through being first in the field with a new discovery, some region will experience a golden age, and it may —as we ourselves have succeeded in doing—develop for a time a system of true democratic institutions.

One of the chief instruments of politics is war, so that it is proper to consider what the future of warfare is likely to be. In this there is a question of the most general importance to be considered first; it is whether the attack or the defence is likely to be the stronger or, putting it figuratively, whether the cavalry or the infantry is to rule the battlefield. The importance of the question may be seen from past history. After the decay of the Roman Empire, the superiority of the cavalry led to more than five hundred years of barbarism in Europe, during which turbulent knights in armour, possessing little merit but a narrow skill in the use of their arms, could hold the world to ransom. They nearly succeeded in destroying the last surviving vestiges of civilization, and it was only later, through the creation of organized armies, helped by the invention of gunpowder, that the infantry again became the predominant arm. After this it became unprofitable to conduct aggressive war in the irresponsible manner that had been profitable earlier, and with this change gradually order and civilization could return to western Europe.

In very recent times there has been a threat that once again the cavalry, in the form of the tank and the aeroplane, might become superior to the infantry. The danger is by no means over, but the experience of the recent war does suggest, rather contrary to expectations, that the infantry still reigns on the battlefield. However, there is a stronger reason which seems to safeguard the future of civilization from destruction by the cavalry. This is that it calls for a very high pitch of civilization to make a tank or an aeroplane. It is by no means unlikely that at some time or other one of the world's provinces may establish itself as a military autocracy and conquer the rest of the world, but to be successful it would have to be—and to stay—at the peak of civilization. So it seems unlikely that, in the future, civilization will be directly destroyed by war, as it was in the Dark Ages.

As to what weapons will be used in war, much will of course depend on how far the fuel problem is solved, but perhaps less than in other fields of activity, because armies and navies always claim, and usually receive, the highest priority in the satisfaction of their demands. It is to be presumed that existing weapons will be improved, that this will be true of both offensive and defensive ones, and that on the whole in the long run the improvements will cancel out. It might be thought natural for me to speculate on the future of the atom bomb, but I shall not do so here, as it is too early to form a critical opinion. Whenever a new weapon is invented, a surge of unreasoning horror goes through the world, which has little relation to the weapon's absolute value. This was so at the time of the introduction of gas-warfare, which military opinion now tends to regard as an inferior weapon, and though the atom bomb must be accepted as far more important, there has not yet been time to assess it properly. The same is true of the various forms of biological warfare that have been considered, and no doubt there will be other wholly new weapons invented from time to time. All these weapons will increase the destructiveness of war, but it must not be forgotten that at the same time there will also be inventions which increase the recuperative power of the defence.

There are two rather different incentives that lead to war. One is fanaticism, the other self-interest. Fanatical wars have been rather rare, and fortunately so, since under the stimulus of a fanatical creed man is ready to inflict, and also to suffer, brutalities to a degree that would hardly be believed possible by those who do not share the creed. It is to be expected that at intervals there will again be such wars. In an over-populated world it is inevitable that there will be a greater callousness about human life, and so it is to be expected that their ferocity will be increased, perhaps even beyond the rather high standard that has been set by the religious wars of the past. Wars stimulated by the milder motive of self-interest may well be more frequent. Here the incentive will often be land-hunger, the wish to find land for a province's starving margin at the expense of another province. In view of the cheapened value of human life there is little likelihood that the hostile population will be treated in a more humane manner than has been the custom in the past, but it will be to the interest of the conqueror to occupy the enemy's land without destroying it. This means that many of the most destructive weapons would not be used, neither the atom bomb which might make the ground uninhabitable for years, nor biological warfare, in any form which might have the effect of making it permanently infertile. This consideration may moderate the evil effects of war to a small extent, but, regarded generally, there is no reason to foresee that war in the future will be any less dreadful than it has been in the past.

CIVILIZATION

It will make a fitting end to my essay to consider the future of civilization; whether it will endure, permanently rising to still greater heights, or whether it is destined to decay after a period of efflorescence, as has happened to so many civilizations in the past. Though we should all agree rather vaguely as to what we mean by civilization, different people may regard very different aspects of it as the central feature. To some it may mean principally great developments in art or literature, to others well-equipped cities and houses, to others a good system of law, to others deep learning, and to others good social conditions. I do not dispute that all or any of these may be involved, but countries could be named, which everyone would concede were civilized, yet which have conspicuously lacked some of these excellences. So for want of a general definition the best way I can describe what seems to me to be involved is by citing an example from the past, the civilization of China.

The Chinese Empire has been civilized for over three thousand years, and until very recent times has enjoyed a very fair measure of isolation. Broadly speaking, during all that time it has retained the same general characteristics. It has been ruled by a succession of dynasties rising and decaying in turn. During the periods of decay, the provinces have often been practically independent, conducting warfare with one another, until at length a new strong hand has arisen to control them. In its forms of government it is true that China seems never to have produced anything like European democracy, but this lack is offset by the creation of a highly organized civil service, not merely centuries but millennia before anything of the kind existed in Europe. All the time the general character of the civilization has been preserved, now in one place, now in another. Sometimes it has been advanced by important new discoveries, such as the invention of printing. All the time there has been a liability to famines, which have killed off millions. The perpetual presence of a margin of starving humanity has set a low value on human life, and has made for callousness in regard to the sufferings of the people. This has led to much cruelty, of a kind we are unfamiliar with now, though it could have been matched anywhere in Europe a few centuries ago. There have been golden ages, when the arts have flourished as nowhere else on earth, and deep learning has been achieved, which we only do not reverence so much as do the Chinese, because it has taken rather a different colour from our own; but even in this we have to concede that the Confucian philosophy has lasted far longer than any of the philosophies of the West. It would seem that in its constancy of character, both in its virtues and in its defects, the Chinese civilization is to be accepted as the model type of a civilization to a greater degree than any of the other civilizations of the world.

In the manner in which it has retained its individual character permanently the Chinese civilization seems pre-eminent, but of course others too have survived for quite long periods. The Roman civilization, though it died in the West, was preserved in a modified form for nearly a thousand years longer in the East. In the same loose sense the Mesopotamian civilization was preserved by the Arabs at Baghdad, until it was overthrown by the Turks, and even so it survived in Egypt and in Spain. There have not been a great many different civilizations in all, so that it is not very safe to generalize; but admitting that some have disappeared leaving no heirs, still the general conclusion must be that in the main there has been at least some survival, if not in the place of origin, then elsewhere. However, that may be, our present civilization is in an incomparably stronger position, for it is dominated by the Scientific Revolution, which, as I have tried to show, makes it basically different from all previous civilizations.

The Scientific Revolution has introduced ways of thinking, which can claim a quality of universality, because they are objective and nearly independent of aesthetic tastes. Even now the community of scientists is quite international, so that they can discuss together the matters that concern them without any thought of national or racial differences. This has never been true of ideas in art, philosophy or religion. For example, the learned of Europe and the learned of China each reverence their own classical literature profoundly, but neither values very highly the classics of the other; whereas in their own subject the scientists of the whole world cannot help valuing the same things. If he is thinking, say, about an electric current, an educated Central African will go through the same processes of thought as an educated Englishman, and no difference in their aesthetic tastes will make any difference between them in this. The Scientific Revolution has changed the world materially in innumerable ways, but perhaps the most important of all is that it has provided a universality in methods of thought that was wanting before. So there is an even stronger reason to believe that the new culture cannot die, than ever held for any of the old civilizations; it has only got to survive in one part of the earth for it to be recoverable everywhere. Even the old civilizations survived for the most part, and it can be regarded as certain that the new culture will be inextinguishable.

A much more difficult question to answer is the question whether civilization will be retained within the same races, or whether there will have to be a perpetual renewal from more barbaric sources. Western Europe, which largely provided the barbarians who recreated the Roman civilization, is itself at the present time in imminent danger of committing suicide. Must civilization always lead to the limitation of families and consequent decay and then replacement from barbaric sources which in turn will go through the same experience? The new developments in birth-control make the threat a great deal more formidable, but in the long run I do not think that it is to be feared. There are already many people with a natural instinctive wish for children, and this wish is sometimes strong enough to outweigh the economic disadvantage which undoubtedly at present attaches to having a family. Such people will tend to have larger families than the rest, and in doing so will at least to some extent hand on the same instinctive wish to a greater number in the next generation. As I have already argued, the limitation of population is an unstable process, which cannot persist. It is very conjectural how long the transformation will take, but as the change that is needed in the balance of human sentiments is very slight, it seems likely that the new balance will not take very long to be established, perhaps thousands of years, but not hundreds of thousands. The first nation or race which can keep its civilization, and at the same time superpose on it this change in the balance of instincts, will have the advantage over all others, both the civilized races that lack the instinct, and the barbarians who have not needed it for their survival. This nation will in consequence dominate the world.

In the establishment of permanently civilized races the most important control will be this small change in the balance of human instincts, because it will have become inherent in the race's nature, and will not need to be taught to each succeeding generation. But it will be helped, and might be much accelerated, if creeds should arise working in the same direction. In the history of mankind creeds will continue to be of very great importance. Among the most important there will always be the creeds, which, without undue fanaticism, inculcate a strong sense of social obligation, since it is only through such creeds that life is possible in crowded communities. There will also no doubt often be fanatical creeds to disturb the peace of the world, and there will be others to comfort the world. I shall not attempt to conjecture what the tenets of these last will be; their main function is to act as a solace to their believers in the very bleak world I have described. It is only this that makes the world tolerable for many people, and this will be much more true in times of real hardship, than in periods of relatively easy prosperity like the present.

The detailed march of history will depend a great deal on the creeds held by the various branches of the human race. It cannot be presumed with any confidence that purely superstitious creeds will always be rejected by civilized communities, in view of the extraordinary credulity shown even now by many reputedly educated people. It is true that there may not be many at the present time, whose actions are guided by an inspection of the entrails of a sacrificial bull, but the progress has not been very great, for there are still many believers in palmistry or astrology. It is to be expected then that in the future, as in the past, there will be superstitions which will notably affect the course of history, and some of them, such as ancestor-worship, will have direct effects on the development of the human species.

But superstitious creeds will hardly be held by the highly intelligent, and it is precisely the creed of these that matters. Is it possible that there should arise a eugenic creed, which—perhaps working through what I have called the method of unconscious selection—should concern itself with the improvement of the inherent nature of man, instead of resting content with merely giving him good but impermanent acquired characters? Without such a creed man's nature will only be changed through the blind operation of natural selection; with it he might aspire to do something towards really changing his destiny.

To conclude, I have cited the past history of China as furnishing the type of an enduring civilization. It seems to provide a model to which the future history of the world may be expected broadly to conform. The scale will of course be altogether vaster, and the variety of happenings cannot by any means be foreseen, but I believe that the underlying ground theme can be foreseen and that in a general way it will be rather like the history of the Chinese Empire. The regions of the world will fall into provinces of ever-changing extent, which most of the time will be competing against one another. Occasionally—more rarely, than has been the case in China—they will be united by some strong arm into an uneasy world-government, which will endure for a period until it falls by the inevitable decay that finally destroys all dynasties. There will be periods when some of the provinces relapse into barbarism, but all the time civilization will survive in some of them. It will survive because it will be based on a single universal culture, derived from the understanding of science; for it is only through this understanding that the multitudes can continue to live. On this basic culture there will be overlaid other cultures, often possessing a greater emotional appeal, which will vary according to climate and race from one province to another. Most of the time and over most of the earth there will be severe pressure from excess populations, and there will be periodic famines. There will be a consequent callousness about the value of the individual's life, and often there will be cruelty to a degree of which we do not willingly think. This however is only one side of the history. On the other side there will be vast stores of learning, far beyond anything we can now imagine, and the intellectual stature of man will rise to ever higher levels. And sometimes new discoveries will for a time relieve the human race from its fears, and there will be golden ages, when man may for a time be free to create wonderful flowerings in science, philosophy and the arts.

XI

EPILOGUE

CAN we do anything about it all? The picture I have
drawn of the future that humanity may expect is
certainly very different from the hopes of the optimistic
idealists of the past and the present. Such people may
argue that many unforeseen wonderful things have happened in the past, and that it is idle to speculate about
what other wonderful things the future may hold in
store. They are forgetting that we are living in an entirely exceptional period, the age of the scientific revolution. I have called it a golden age, and I would remind
them that during the course of history man has assigned
the epoch of the golden age at least as often to the past
as to the future.

Anyone who disagrees with my forecast must try to
get beyond a vague optimism, which merely expresses
the confidence that "something will turn up". In particular he must find a really solid reason which shows how
the threat of over-population will be avoided; the observation that it has been avoided in some countries during
the last few years is not enough. Let him then give the
fullest rein to his imagination, let him suppose that anything is permissible, but let him follow out the consequences to their conclusion. I will venture to say that if
he does so he will find that one or other of two alter-

natives is the result. Either he will come to general conclusions not so very different from mine; he will find that his utopia, however pleasant it may be in other ways, in the long run will suffer from many disagreeable features of the kind that I have been considering. Or else he will find that his imagination has gone so far out of the realms of reality that it contradicts the physical or the biological laws of nature.

Nevertheless for all of us it is intolerable to think of the future unfolding itself in complete predestined inevitability for the eternity of a million years. There are two things we must do; one is to know, the other to act. As to knowing, in my introductory chapter I described an analogy in mechanics, and I suggested that it should be possible to discover a set of laws, like the laws of thermodynamics, which would place absolute limits on what can be done by humanity. Biological laws cannot be expected to have the same hard outline as physical laws, but still there are absolute laws limiting what an animal can do, and similar laws will limit man not only on his physical side, but also on his intellectual side. If these could be clearly stated, we should recognize that many attempts that have been made at improving man's estate were hopeless.

It is for others, better versed than I am in the biological sciences, to work out these laws, and it is in all humility that I put forward the basis, on which, it may be, that they could be founded. The first principle is that man, as an animal, obeys the law of variation of species, which condemns human nature to stay nearly constant for a million years. The perfectibility of mankind, the aim of so many noble spirits, is foredoomed by this principle. The second is that man is a wild animal, and that doctrines drawn from the observation of domestic animals are quite inapplicable to him. The third principle is the non-inheritance of acquired characters, a principle familiar in animal biology, but all too seldom invoked in connection with human beings. If these, and any further principles as well, or any alternatives to them, were accepted, it might sometimes be possible through them to show up the absurdities of bad statesmanship, and certainly it would be the part of a wise statesman to work within their limitations, because only so could he hope to achieve success.

What action can be taken about the future of the human race? I am afraid that the answer must be very little indeed, and this is for the simple reason that most human beings do not care in the least about the distant future. Most care about the conditions that will affect their children and their grandchildren, but beyond that the situation seems too unreal, and even for those who do think about the more distant future, the uncertainties are too great to suggest any clear course of action. For example, consider the inevitable fuel shortage that is to come so soon. I know that my sons will not suffer from it very seriously, and I know that the fifteenth generation of my descendants will get no coal at all. Am I likely to refrain from putting coal on the fire on a cold evening by the thought that it may make one of my fourteenth descendants suffer for it? Such matters are so unreal to our minds, that it is not to be expected that they will ever be given much weight. Life is always precarious, and it is so hard to be sure of keeping alive for even ten years, that it is not surprising that no one should care much about what is going to happen even as short a time ahead as a century. In hardly any of the affairs of the world will man really be interested in the more distant future.

Still for the sake of the distant future something can be attempted more profitable than has been usual hitherto. Attempts at improving the lot of mankind have all hitherto been directed towards improving his conditions, but not his nature, and as soon as the conditions lapse all is lost. The only hope is to use our knowledge of biology in such a way that all would not be lost with the lapse of the conditions. The principles of heredity offer an anchor which will permanently fix any gains that there may be in the quality of mankind.

In final conclusion I had better declare my personal inclination. I do care very much about the future of the world, and I want most intensely my own descendants to play their part in it. However bleak the future, I am not content with the thought that it should be a world in which I have had no continuing part. No matter whether in the long ages to come life is to be a joy or a misery—and certainly much of it will be a misery—it will be an adventure that is well worth while.

INDEX

INDEX